DUSHKIN STUDENT ATLAS of ECONOMIC DEVELOPMENT

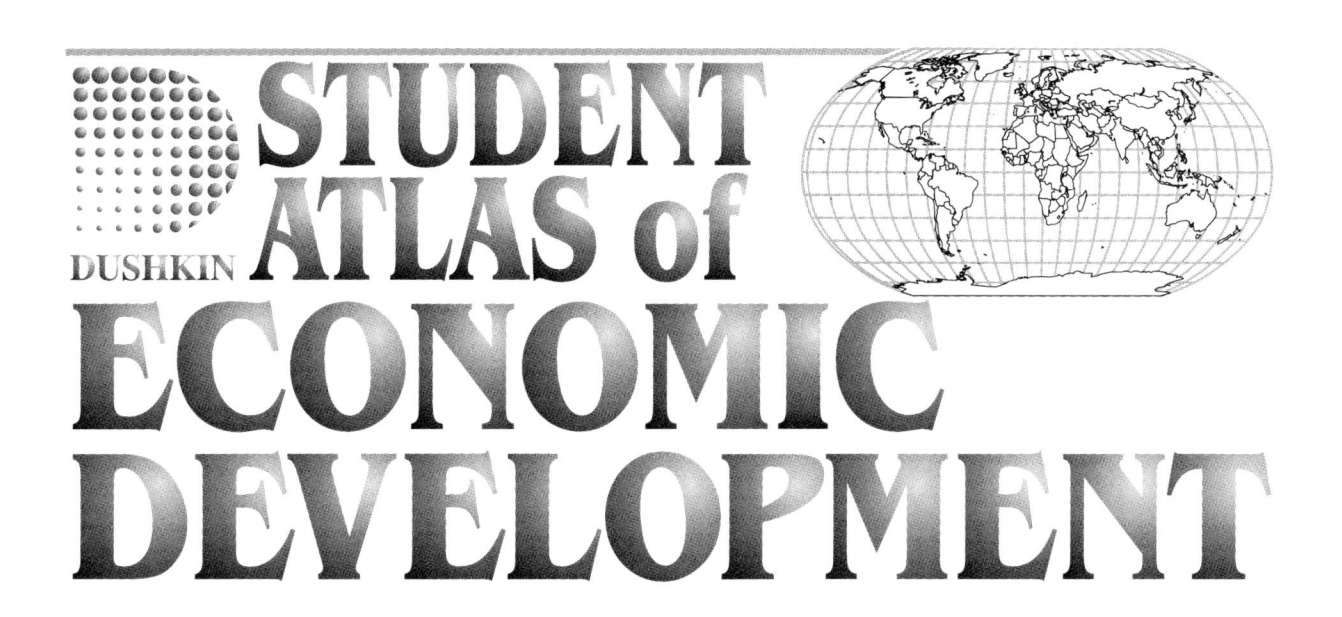

DUSHKIN STUDENT ATLAS of ECONOMIC DEVELOPMENT

John L. Allen
University of Connecticut

Dushkin/McGraw·Hill
A Division of The McGraw·Hill Companies

Book Team

Vice President & Publisher *Jeffrey L. Hahn*
Managing Editor *John S. L. Holland*
Production Manager *Brenda S. Filley*
Editor *Pamela Carley*
Copy Editor *Joshua Safran*
Designer *Charles Vitelli*
Typesetting Supervisor *Juliana Arbo*
Proofreader *Diane Barker*
Graphics *Shawn Callahan*
Marketing Manager *Mike Matera*

Dushkin/McGraw·Hill
A Division of The McGraw·Hill Companies

The credit section for this book begins on page 101 and is considered an extension of the copyright page.

Cover design *Shawn Callahan*
Cover photo courtesy of NASA

Copyright © 1997 The McGraw-Hill Companies, Inc.
All rights reserved

Library of Congress Catalog Card Number: 96-71801

ISBN 0-697-36515-8

No part of this publication may be reproduced, stored in a retrieval system, or transmitted, in any form or by any means, electronic, mechanical, photocopying, recording, or otherwise, without the prior written permission of the publisher.

Printed in the United States of America

10 9 8 7 6 5 4 3 2 1

A Note to the Student

Economic development remains the great challenge of our time. In spite of remarkable (and often unnoticed) progress since World War II in reducing child mortality, increasing life expectancy, eradicating disease, and increasing per capita income, more than 20 percent of the world's population still lives in abject poverty. Millions die each year from diseases brought on by inadequate diets, pollution of air and water, substandard housing, and poor or absent medical services. The maps in this atlas serve to illustrate the dimensions of both the fact and the promise of economic development and can help you, as a student, to understand the vast and complex drama of international economic development. Use this atlas in conjunction with your other course materials and it will help you become more knowledgeable of this crucial global process.

The maps and data sets in the *DPG Student Atlas of Economic Development* are designed to introduce you to the importance of the connections between geography and world economics. The maps are not perfect representations of reality—no maps ever are—but they do represent "models," or approximations of the real world, that should aid in your understanding of global economic development.

A Word about Data Sources

At the very outset of your study of this atlas, you should be aware of some limitations of the maps and data tables. In some instances, a map or a table may have missing data. This may be the result of the failure of a country to report information to a central international body (such as the United Nations or the World Bank). Alternatively, it may reflect shifts in political boundaries, internal conflicts, or changes in responsibility for reporting data that have caused certain countries (such as those that made up the former Soviet Union or the former Yugoslavia) to delay their reports. It is always our wish to be as up-to-date as possible; subsequent editions of this atlas will probably have increased data on countries like Slovenia, Ukraine, or Uzbekistan. In the meantime, as events continue to restructure our world, it's an exciting time to be a student of international economics!

You will find your study of this atlas more productive in relation to your exploration of economic development if you examine the maps on the following pages in terms of five distinct analytical themes:

1. *Location: Where Is It?* This involves a focus on the precise location of places in both *absolute* terms (the latitude and longitude of a place) and *relative* terms (the location of a place in relation to the location of other places). When you think of location, you should automatically think of both forms. Knowing something about absolute location will help you to understand a variety of features of physical geography, since such key elements are so closely related to their position on the earth. But it is equally important to think of location in relative terms. Where places are located in relation to other places is often more important as a determinant of social, economic, and cultural characteristics than are the factors of physical geography.
2. *Place: What Is It Like?* This encompasses the economic, political, cultural, environmental, and other characteristics that give a place its identity. You should seek to understand the similarities and differences between places by exploring their basic characteristics. Why are some places with similar

environmental characteristics so very different in economic, cultural, social, and political ways? Why are other places with such different environmental characteristics so seemingly alike in terms of their institutions, their economies, and their cultures?

3. *Human/Environment Interactions: How Is the Landscape Shaped?* This focuses on the ways in which people respond to and modify their environments. On the world stage, humans are not the only part of the action. The environment also plays a role in the drama of economic development. But the characteristics of the environment do not exert a controlling influence over human activities; they only provide a set of alternatives from which different cultures, in different times, make their choices. Observe the relationship between the basic elements of physical geography such as climate and terrain and the host of ways in which humans have used the land surfaces of the world.

4. *Movement: How Do People Stay in Touch?* This examines the transportation and communication systems that link people and places. Movement or "spatial interaction" is the chief mechanism for the spread of ideas and innovations from one place to another. It is spatial interaction that validates the old cliché, "the world is getting smaller." We find McDonald's restaurants in Tokyo and Honda automobiles in New York City because of spatial interaction. Advanced transportation and communication systems have transformed the world in which your parents were born. And the world that greets your children will be very different from your world. None of these changes would happen without the force of movement or spatial interaction.

5. *Regions: Worlds within a World*. This theme, perhaps the most important for this atlas, helps to organize knowledge about the land and its people. The world consists of a mosaic of "regions" or areas that are somehow different and distinctive from other areas. The region of Anglo-America (the United States and Canada) is, for example, different enough from the region of Western Europe that geographers clearly identify them as two unique and separate areas. Yet despite their differences, Anglo-Americans and Europeans share a number of similarities: common cultural backgrounds, comparable economic patterns, and shared religious traditions. Conversely, although the regions of Anglo-America and Eastern Asia are also easily distinguished as distinctive units of the earth's surface, they have a number of shared physical environmental characteristics. But those who live in Eastern Asia and Anglo-America have fewer similarities and more differences between them than do residents of Western Europe and Anglo-America; Eastern Asians and Anglo-Americans have different cultural traditions, different institutions, and different linguistic and religious patterns. An understanding of both the differences and similarities between regions like Anglo-America and Western Europe, on the one hand, and Anglo-America and Eastern Asia, on the other, will help you to understand much that has happened in the human past or that is currently transpiring in the world around you. At the very least, an understanding of regional similarities and differences will help you to interpret what you read on the front page of your daily newspaper or view on the evening news report on your television set.

Not all of these concepts will be immediately apparent on each of the 45 maps and 13 tables in this atlas. But if you study the contents of *Atlas of Economic Development* along with the reading of your text and think about the five themes, the maps and tables and the text will complement one another and improve your understanding of global economic development.

John L. Allen

Acknowledgments

The quality of this book was improved by the insights provided by the following reviewers:

Don Cole, Drew University
Matthew Marlin, Duquesne University
Susan M. Randolph, University of Connecticut
Howard Stein, Roosevelt University
Irvin Weintraub, Towson State University

Table of Contents

To the Student   v
Acknowledgements   vii

Part I. The Geographic Factor: Global Patterns   1

Map 1 Current World Political Boundaries   2
Map 2 World Climate Regions   4
Map 3 World Topography   6
Map 4 World Ecological Regions   8
Map 5 World Natural Hazards   10
Map 6 Land Use Patterns of the World   12
Map 7 World Population Density   14
Table A World Countries: Area, Population, and Population Density, 1996   16

Part II. The Human Factor: The World's People   19

Map 8 Population Growth Rates   20
Map 9 Total Fertility Rate   21
Map 10 Infant Mortality Rate   22
Map 11 Average Life Expectancy at Birth   23
Map 12 Population by Age Group   24
Map 13 World Daily Per Capita Food Supply   25
Map 14 Child Malnutrition   26
Map 15 Illiteracy Rate   27
Map 16 Primary School Enrollment   28
Map 17 Total Labor Force   29
Map 18 Female Labor Force   30
Map 19 Female/Male Inequality in Education and Employment   31
Map 20 Urban Population   32
Map 21 Countries with Significant Minority Group Populations   33
Map 22 Quality of Life: The Human Development Index   34
Table B Size and Growth of Population, 1950-2025   35
Table C World Countries: Demography, 1975-1995   39
Table D Mortality, Health, and Nutrition, 1970-1995   43
Table E Education and Literacy, 1970-1993   47

Part III. The Economic Factor: The World's Economies   51

Map 23 Rich and Poor Countries: Gross National Product   52
Map 24 Gross National Product Per Capita   53
Map 25 Economic Growth: Change in Gross National Product, 1983-1993   54
Map 26 Employment by Sector   55
Map 27 Gross Domestic Product—Share in Agriculture   56

Map 28 Production of Staples—Cereals, Roots, and Tubers   57
Map 29 Agricultural Production Per Capita   58
Map 30 Gross Domestic Product—Share in Exports   59
Map 31 Exports of Primary Products   60
Map 32 Dependence on Trade   61
Map 33 Gross Domestic Product—Share in Investment   62
Map 34 Relative Wealth of Nations: Purchasing Power Parity   63
Map 35 Central Government Expenditures Per Capita   64
Map 36 Energy Production   65
Map 37 Energy Requirements Per Capita   66
Map 38 Production of Crucial Materials   67
Map 39 Composition of Crucial Materials   68
Table F World Countries: Basic Economic Indicators   69
Table G World Countries: Agricultural Operations, 1993   73
Table H World Countries: Energy Production, Consumption, and Requirements, 1993   76
Table I World Countries: Infrastructure   79

Part IV. The Environmental Factor: World Environmental Conditions   83

Map 40 Reliance on Traditional Energy Sources: Per Capita Use of Biomass Energy   84
Map 41 Deforestation and Desertification   85
Map 42 Soil Degradation   86
Map 43 Annual Water Use   87
Map 44 Air and Water Quality   88
Map 45 Per Capita CO_2 Emissions   89
Table J Land Use   90
Table K Human-Induced Soil Destruction, 1945–1990   94
Table L Water Resources, 1995   95
Table M Greenhouse Gas Emissions, 1992   98

Sources   101

Part I

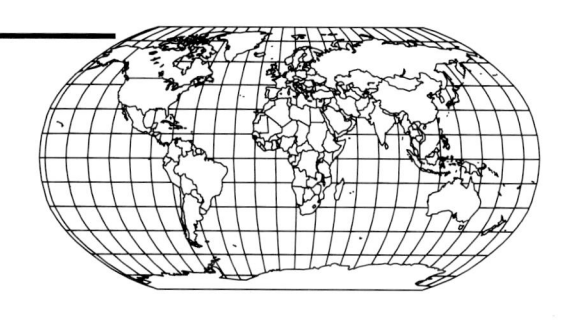

The Geographic Factor: Global Patterns

Map 1 Current World Political Boundaries

Map 2 World Climate Regions

Map 3 World Topography

Map 4 World Ecological Regions

Map 5 World Natural Hazards

Map 6 Land Use Patterns of the World

Map 7 World Population Density

Table A World Countries: Area, Population, and Population Density, 1996

Map 1 Current World Political Boundaries

Although economic regions should, in theory, have no political base, the most important components of the international economic system are still individual countries or "states." The boundaries of countries are the basis of political division in the world, and for most people nationalism is the strongest source of both political and economic identification.

Scale: 1 to 125,000,000

Note: All world maps are Robinson projection.

NORTH SEA
NORWAY
SWEDEN
BALTIC SEA
ESTONIA
0 100 Miles
0 100 Kilometers
DENMARK
LATVIA
RUSSIA
LITHUANIA
RUSSIA
NETHERLANDS
BELARUS
GERMANY
POLAND
BELGIUM
CZECH REPUBLIC
UKRAINE
LUXEMBOURG
SLOVAKIA
FRANCE
LIECHTENSTEIN
MOLDOVA
SWITZERLAND
AUSTRIA
HUNGARY
ROMANIA
SLOVENIA
YUGOSLAVIA (SERBIA-MONTENEGRO)
CROATIA
SAN MARINO
BOSNIA-HERZEGOVINA
BLACK SEA
MONACO
ITALY
BULGARIA
ALBANIA
MACEDONIA
GREECE
TURKEY
MEDITERRANEAN SEA
MALTA
ARCTIC OCEAN
Arctic Circle
ICELAND
NORWAY
SWEDEN
FINLAND
UNITED KINGDOM
IRELAND
FRANCE
ANDORRA
SPAIN
PORTUGAL
TURKEY
R U S S I A
KAZAKHSTAN
MONGOLIA
UZBEKISTAN
KYRGYZSTAN
TURKMENISTAN
TAJIKISTAN
NORTH KOREA
SOUTH KOREA
JAPAN
NORTH PACIFIC OCEAN
C H I N A
TUNISIA
CYPRUS
LEBANON
SYRIA
IRAQ
IRAN
AFGHANISTAN
MOROCCO
ISRAEL
JORDAN
KUWAIT
QATAR
BAHRAIN
PAKISTAN
NEPAL
BHUTAN
ALGERIA
LIBYA
EGYPT
SAUDI ARABIA
MYANMAR (BURMA)
Tropic of Cancer
I N D I A
TAIWAN
MAURITANIA
MALI
NIGER
ERITREA
UNITED ARAB EMIRATES
OMAN
YEMEN
BANGLADESH
LAOS
CHAD
SUDAN
THAILAND
VIETNAM
PHILIPPINES
MARSHALL ISLANDS
NIGERIA
DJIBOUTI
ETHIOPIA
SRI LANKA
CAMBODIA (KAMPUCHEA)
PALAU
UGANDA
SOMALIA
BRUNEI
MICRONESIA
CAMEROON
MALDIVES
MALAYSIA
CENTRAL AFRICAN REPUBLIC
RWANDA
Equator
SINGAPORE
I N D O N E S I A
NAURU
KIRIBATI
SÃO TOMÉ AND PRÍNCIPE
BURUNDI
KENYA
EQUATORIAL GUINEA
ZAIRE
SEYCHELLES
PAPUA NEW GUINEA
SOLOMON ISLANDS
GABON
TANZANIA
MALAWI
TUVALU
CONGO
COMOROS
MOZAMBIQUE
INDIAN OCEAN
ANGOLA
ZAMBIA
VANUATU
FIJI
MADAGASCAR
NAMIBIA
MAURITIUS
Tropic of Capricorn
BOTSWANA
AUSTRALIA
ZIMBABWE
SWAZILAND
SOUTH AFRICA
LESOTHO
NEW ZEALAND
MAURITANIA
MALI
SENEGAL
NIGER
GAMBIA
GUINEA-BISSAU
BURKINA FASO
GUINEA
BENIN
SIERRA LEONE
ATLANTIC OCEAN
NIGERIA
IVORY COAST
GHANA
LIBERIA
0 150 300 Miles
0 150 300 Kilometers
TOGO
0 100 Miles
0 100 Kilometers
CASPIAN SEA
RUSSIA
BLACK SEA
GEORGIA
AZERBAIJAN
ARMENIA
TURKEY
AZERBAIJAN
IRAN

Map 2 World Climate Regions

Climates of the World

Tropical Moist Climates
- Rainforest
- Monsoon
- Savanna

Dry Climates
- Warm Desert
- Warm Steppe
- Cool Desert
- Cool Steppe

Midlatitude Climates
- Summer Dry or Mediterranean
- Moist Subtropical
- Marine West Coast
- Cool Forest
- Subarctic

Polar Climates
- Tundra
- Ice Cap

Azonal Climates
- Undifferentiated Highlands

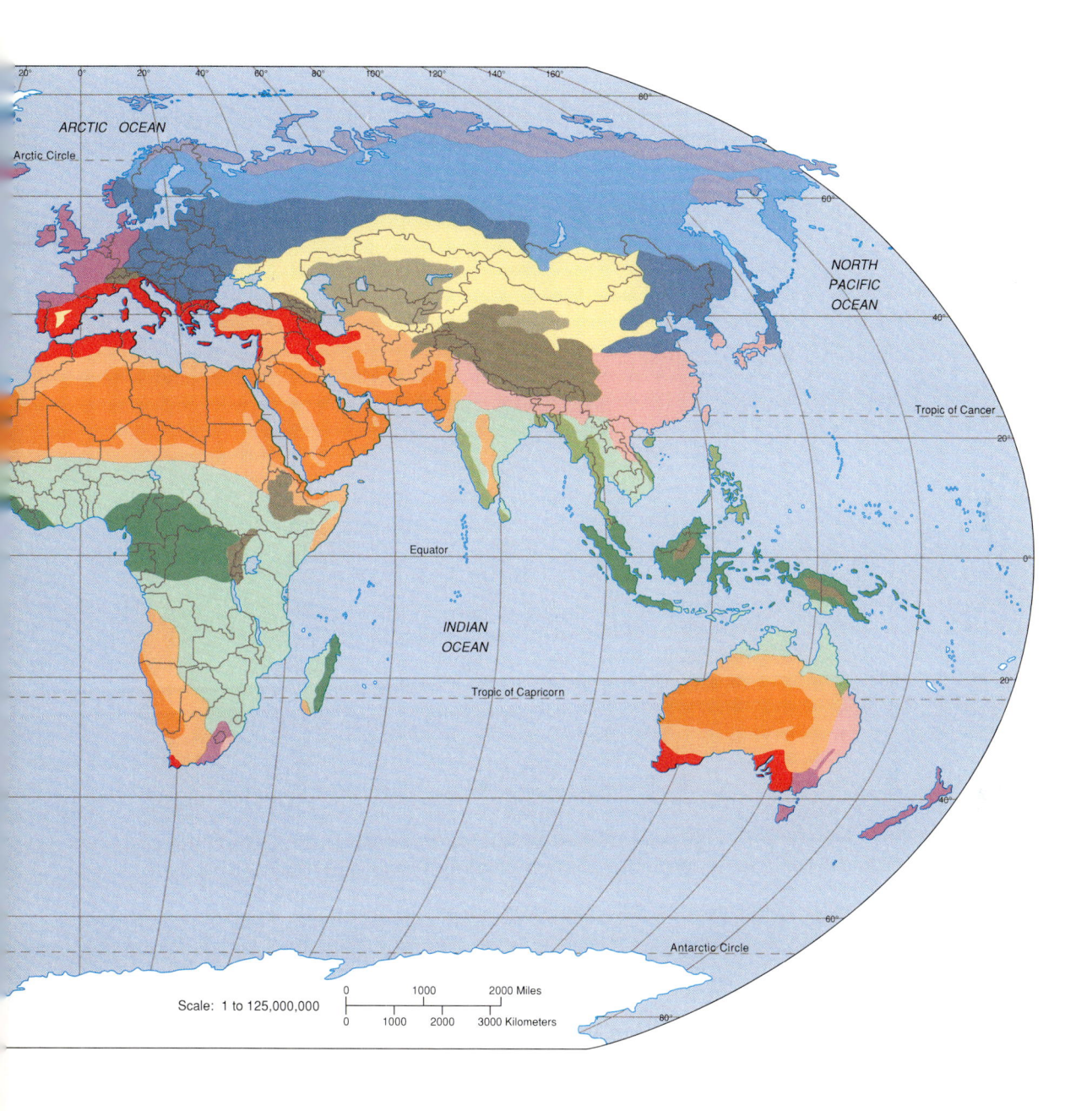

Of the world's many physical geographic features, climate (the long-term average of such weather conditions as temperature and precipitation) is the most important. It is climate that conditions the types of natural vegetation patterns and the types of soil that will exist in an area. It is also climate that determines the availability of our most precious resource: water. From an economic standpoint, the world's most important activity is agriculture; no other element of physical geography is more important for agriculture than climate.

Map 3 World Topography

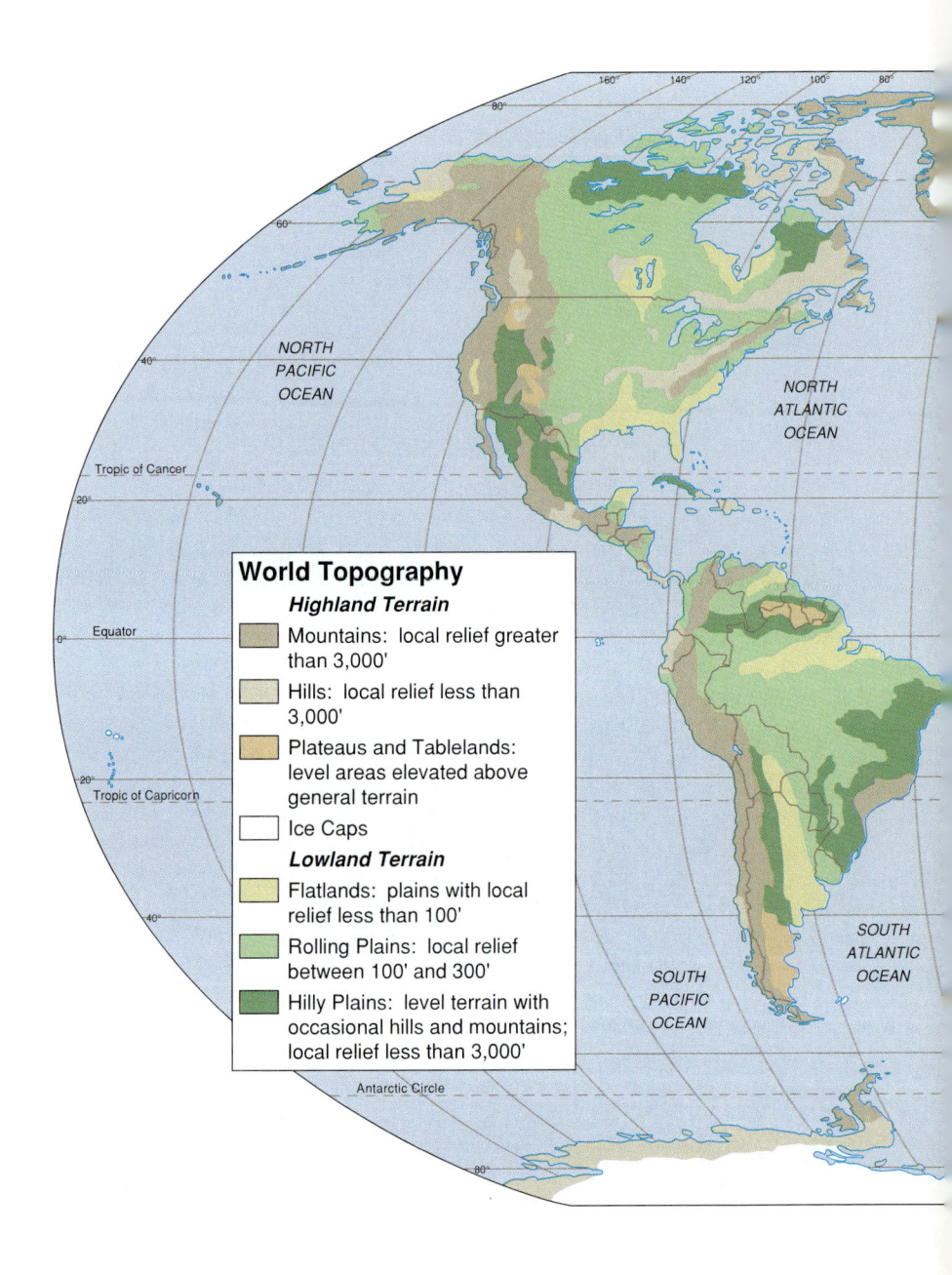

World Topography

Highland Terrain
- Mountains: local relief greater than 3,000'
- Hills: local relief less than 3,000'
- Plateaus and Tablelands: level areas elevated above general terrain
- Ice Caps

Lowland Terrain
- Flatlands: plains with local relief less than 100'
- Rolling Plains: local relief between 100' and 300'
- Hilly Plains: level terrain with occasional hills and mountains; local relief less than 3,000'

Second only to climate as a conditioner of human activity—particularly in agriculture and also in the location of cities and industry—is topography or terrain. It is what we often call "landforms." A comparison of this map with the map of land use (Map 6) will show that most of the world's productive agricultural zones are located in lowland regions. Where large regions of agricultural productivity are found, we also tend to find urban concentrations and, with cities, we find industry. There is also a good spatial correlation between this map of landforms and the map showing the distribution and density of the human population (Map 7). The landforms shown on this map are primarily the result of extremely gradual geologic activity, such as the long-term movement of crustal plates (sometimes called "continental drift"). This activity occurs over hundreds of millions of years. Also important is the more rapid (but still slow by human standards) erosional activity of water, wind, glacial ice, and waves, tides, and currents. Significant erosional changes occur over a span of a few million years. Some landforms may be produced by abrupt or "cataclysmic" events such as a major volcanic eruption or a meteor strike, but these events are relatively rare and their effects are usually too minor to show up on a map of this scale.

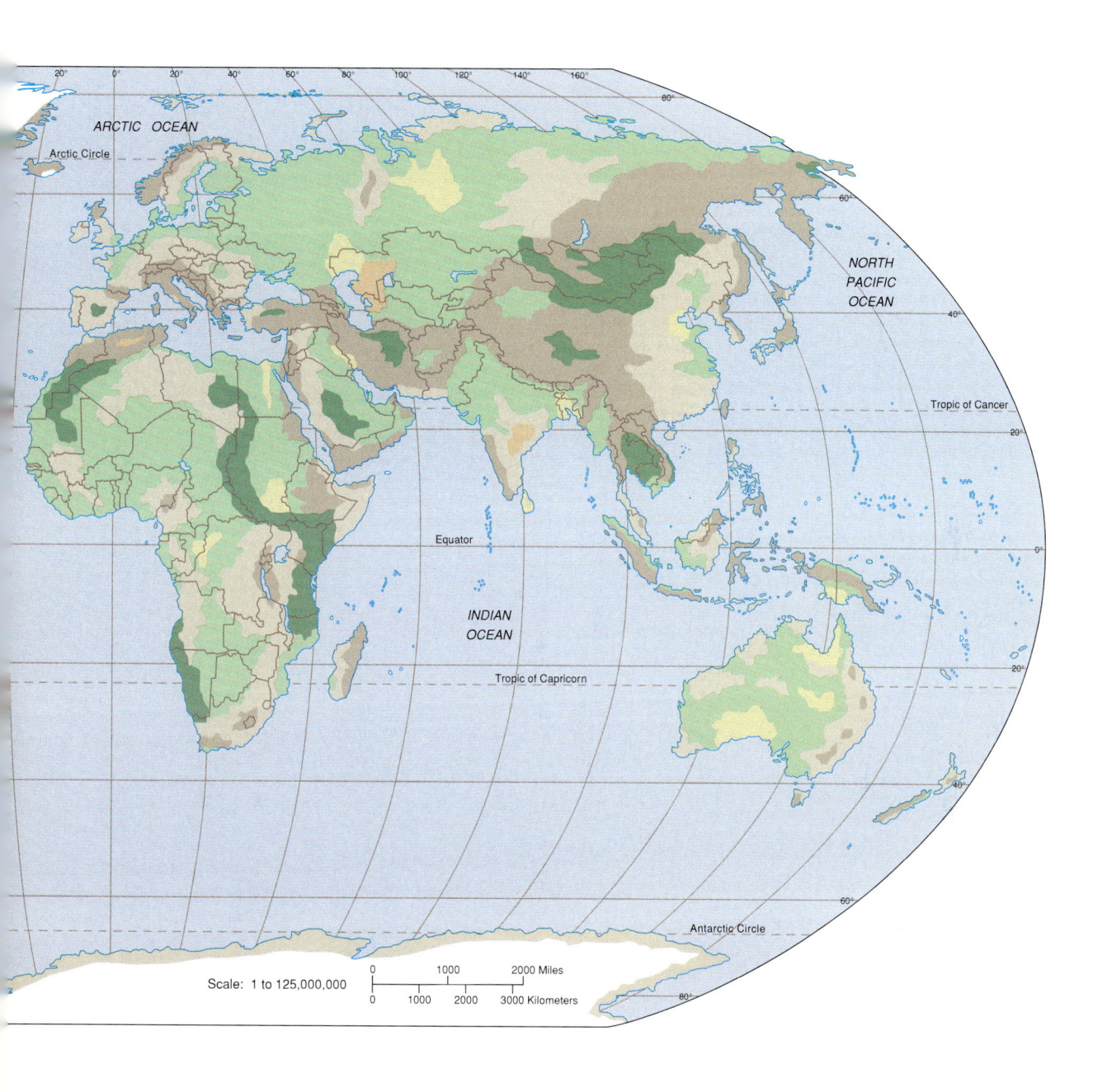

Map 4 World Ecological Regions

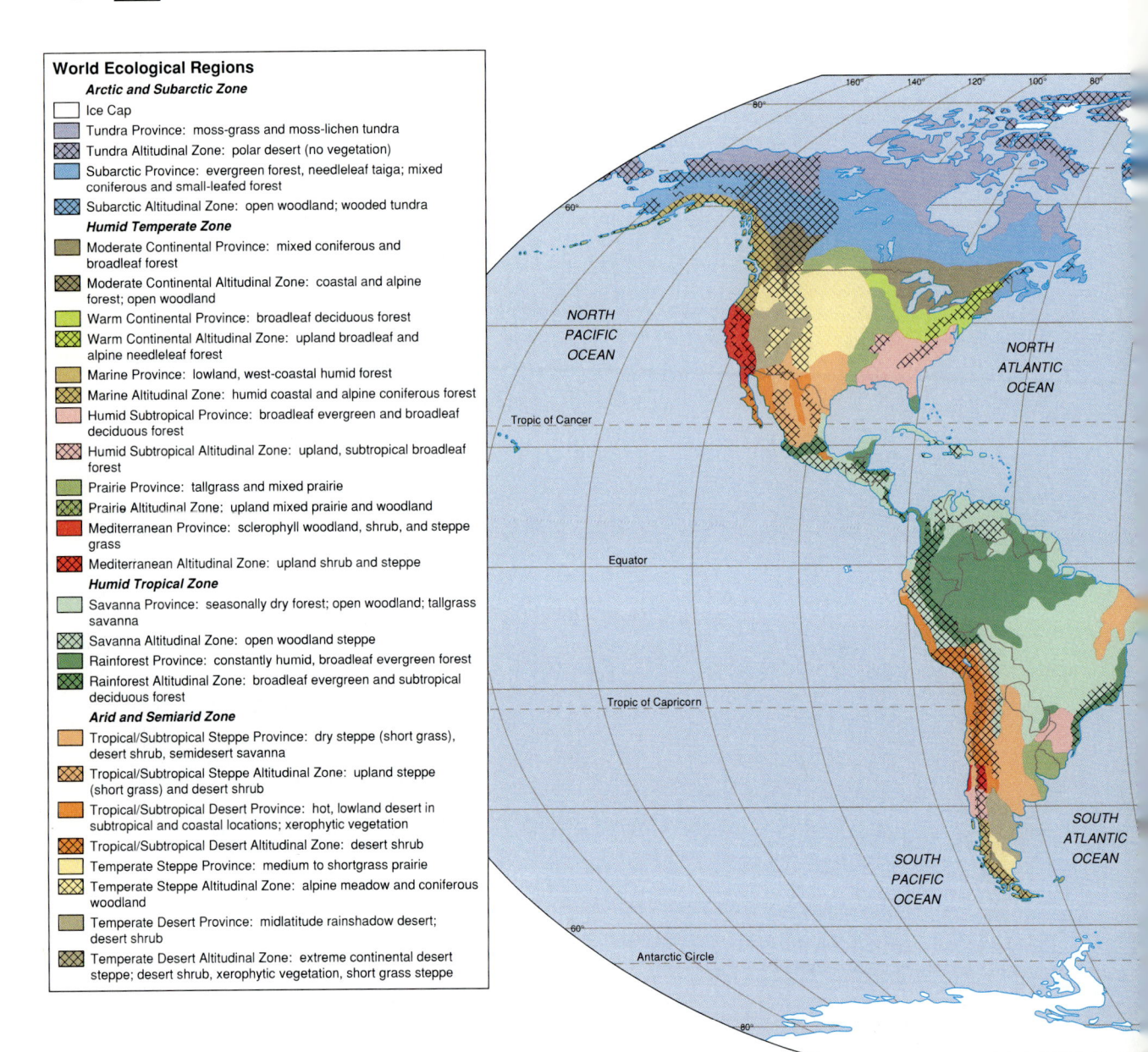

World Ecological Regions

Arctic and Subarctic Zone
- Ice Cap
- Tundra Province: moss-grass and moss-lichen tundra
- Tundra Altitudinal Zone: polar desert (no vegetation)
- Subarctic Province: evergreen forest, needleleaf taiga; mixed coniferous and small-leafed forest
- Subarctic Altitudinal Zone: open woodland; wooded tundra

Humid Temperate Zone
- Moderate Continental Province: mixed coniferous and broadleaf forest
- Moderate Continental Altitudinal Zone: coastal and alpine forest; open woodland
- Warm Continental Province: broadleaf deciduous forest
- Warm Continental Altitudinal Zone: upland broadleaf and alpine needleleaf forest
- Marine Province: lowland, west-coastal humid forest
- Marine Altitudinal Zone: humid coastal and alpine coniferous forest
- Humid Subtropical Province: broadleaf evergreen and broadleaf deciduous forest
- Humid Subtropical Altitudinal Zone: upland, subtropical broadleaf forest
- Prairie Province: tallgrass and mixed prairie
- Prairie Altitudinal Zone: upland mixed prairie and woodland
- Mediterranean Province: sclerophyll woodland, shrub, and steppe grass
- Mediterranean Altitudinal Zone: upland shrub and steppe

Humid Tropical Zone
- Savanna Province: seasonally dry forest; open woodland; tallgrass savanna
- Savanna Altitudinal Zone: open woodland steppe
- Rainforest Province: constantly humid, broadleaf evergreen forest
- Rainforest Altitudinal Zone: broadleaf evergreen and subtropical deciduous forest

Arid and Semiarid Zone
- Tropical/Subtropical Steppe Province: dry steppe (short grass), desert shrub, semidesert savanna
- Tropical/Subtropical Steppe Altitudinal Zone: upland steppe (short grass) and desert shrub
- Tropical/Subtropical Desert Province: hot, lowland desert in subtropical and coastal locations; xerophytic vegetation
- Tropical/Subtropical Desert Altitudinal Zone: desert shrub
- Temperate Steppe Province: medium to shortgrass prairie
- Temperate Steppe Altitudinal Zone: alpine meadow and coniferous woodland
- Temperate Desert Province: midlatitude rainshadow desert; desert shrub
- Temperate Desert Altitudinal Zone: extreme continental desert steppe; desert shrub, xerophytic vegetation, short grass steppe

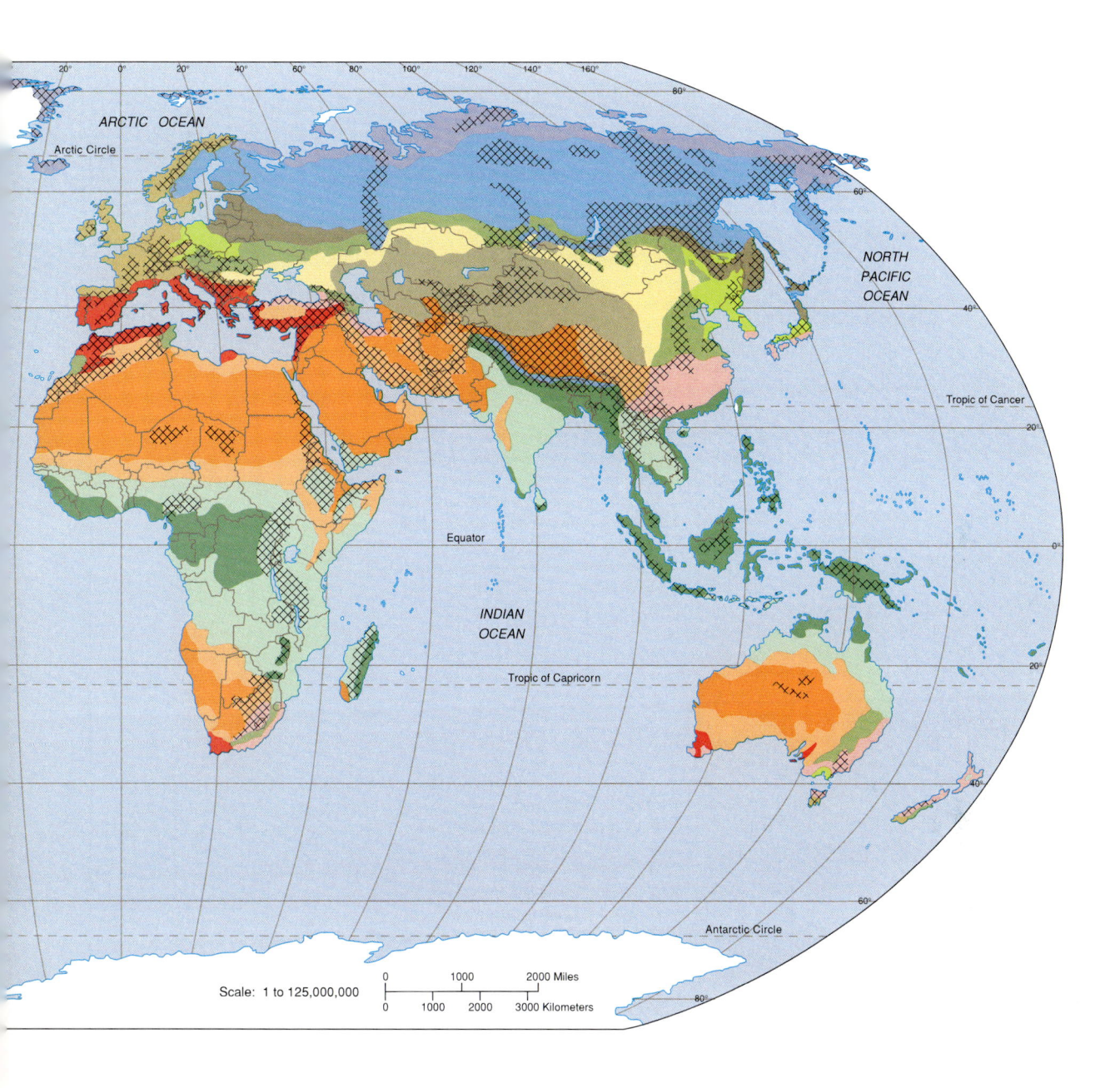

Ecology is the study of the relationships between living organisms and their environmental surroundings. Ecological regions are distinctive areas within which unique sets of organisms and environments are found. Within each ecological region, a particular combination of vegetation, wildlife, soil, water, climate, and terrain defines that region's "habitability," or ability to support life, including human life. Like climate and landforms, ecological relationships are crucial to the existence of agriculture, the most basic of our economic activities, and important for most other kinds of economic activity as well.

Map 5 World Natural Hazards

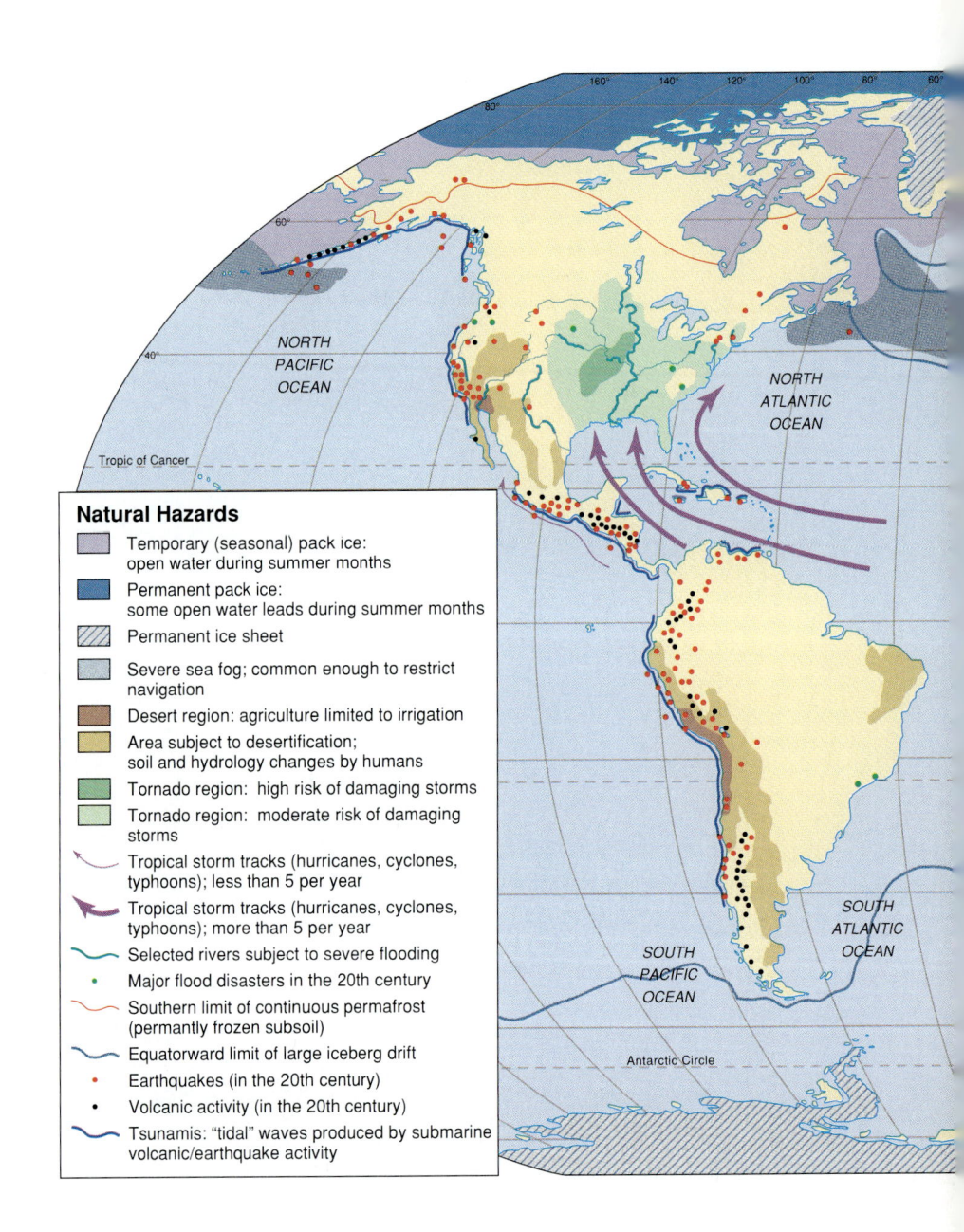

Natural Hazards

- Temporary (seasonal) pack ice: open water during summer months
- Permanent pack ice: some open water leads during summer months
- Permanent ice sheet
- Severe sea fog; common enough to restrict navigation
- Desert region: agriculture limited to irrigation
- Area subject to desertification; soil and hydrology changes by humans
- Tornado region: high risk of damaging storms
- Tornado region: moderate risk of damaging storms
- Tropical storm tracks (hurricanes, cyclones, typhoons); less than 5 per year
- Tropical storm tracks (hurricanes, cyclones, typhoons); more than 5 per year
- Selected rivers subject to severe flooding
- Major flood disasters in the 20th century
- Southern limit of continuous permafrost (permanently frozen subsoil)
- Equatorward limit of large iceberg drift
- Earthquakes (in the 20th century)
- Volcanic activity (in the 20th century)
- Tsunamis: "tidal" waves produced by submarine volcanic/earthquake activity

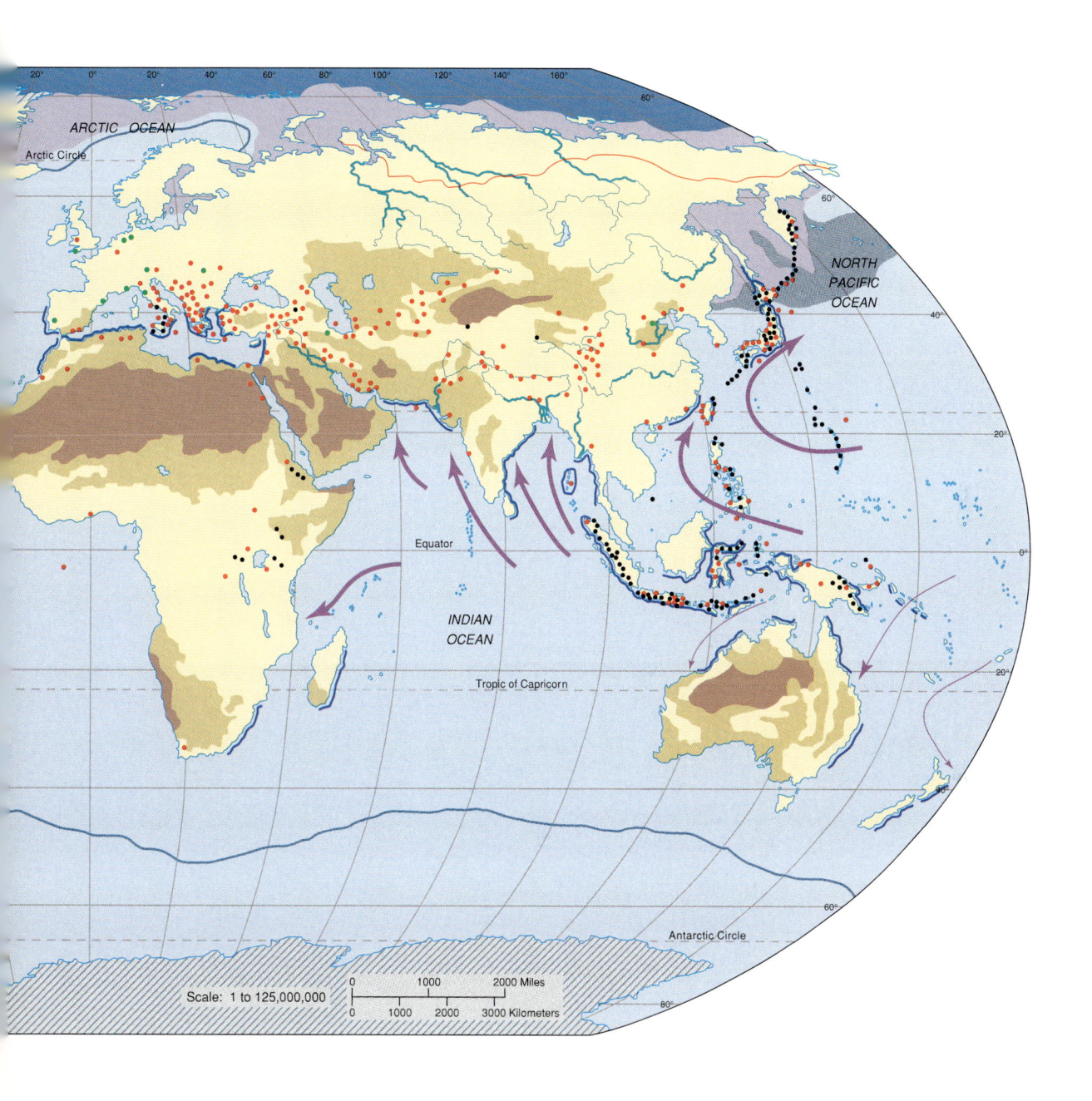

Unlike other elements of physical geography, natural hazards are unpredictable. However, there are certain regions where the *probability* of the occurrence of a particular natural hazard is high. This map shows regions affected by major natural hazards at rates that are higher than the global norm. Persistent natural hazards may undermine the utility of an area for economic purposes.

Map 6 Land Use Patterns of the World

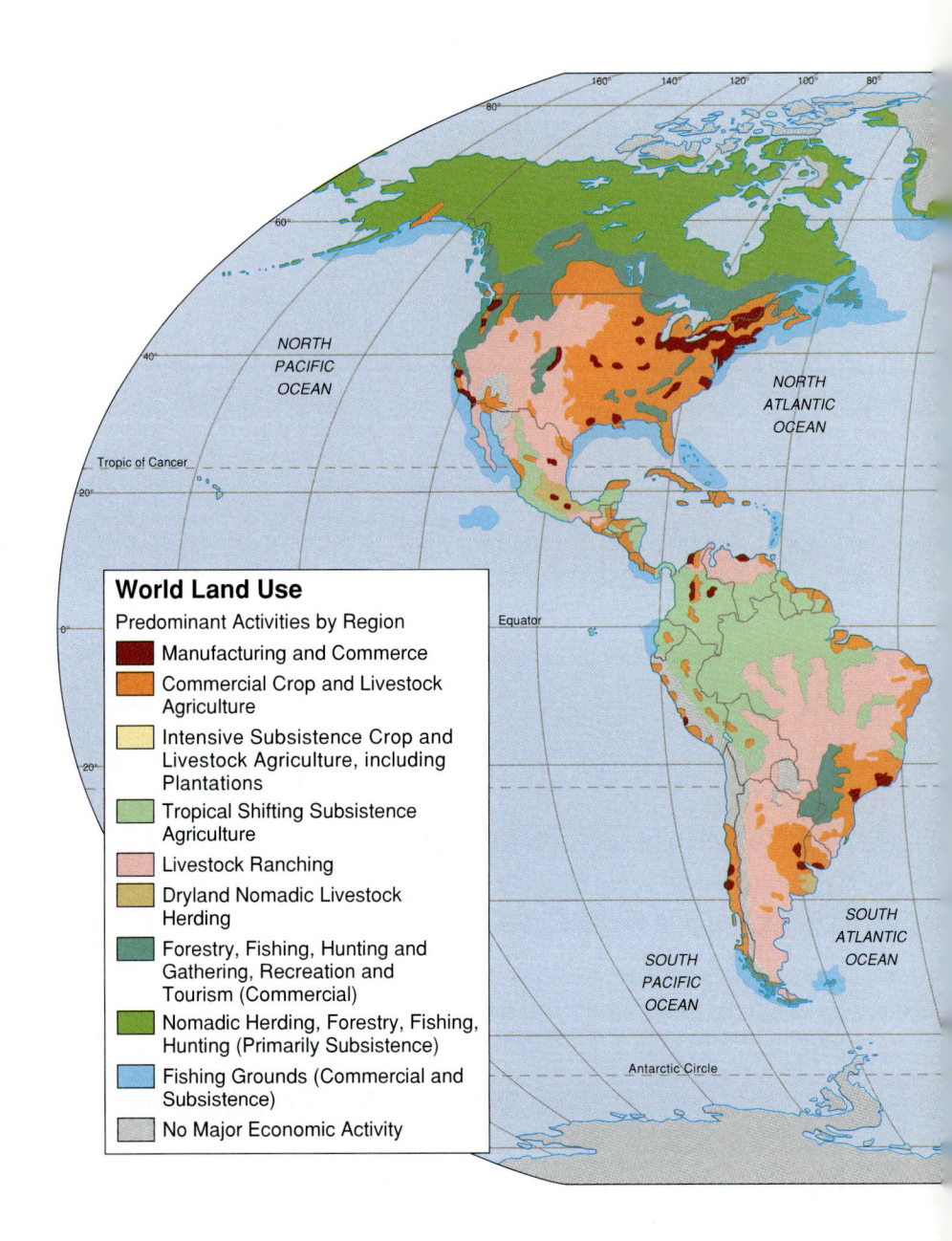

World Land Use
Predominant Activities by Region

- Manufacturing and Commerce
- Commercial Crop and Livestock Agriculture
- Intensive Subsistence Crop and Livestock Agriculture, including Plantations
- Tropical Shifting Subsistence Agriculture
- Livestock Ranching
- Dryland Nomadic Livestock Herding
- Forestry, Fishing, Hunting and Gathering, Recreation and Tourism (Commercial)
- Nomadic Herding, Forestry, Fishing, Hunting (Primarily Subsistence)
- Fishing Grounds (Commercial and Subsistence)
- No Major Economic Activity

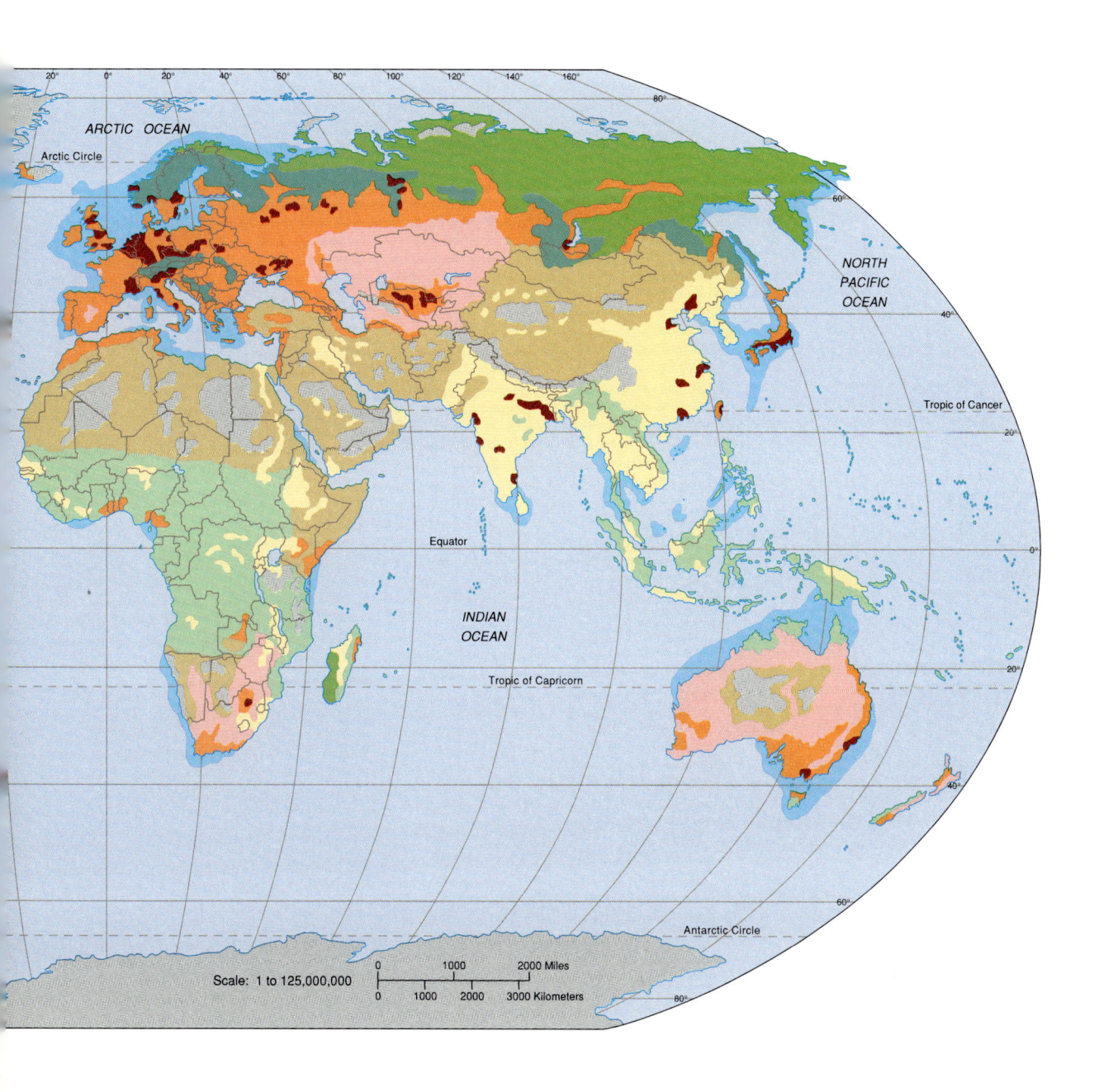

Many of the major land use patterns of the world (such as urbanization, industry, and transportation) cover a relatively small area and are not easily seen on maps, but the most important uses that people make of the earth's surface have more far-reaching effects. This map illustrates, in particular, the variations in primary land uses (such as agriculture) for the entire world. Note particularly the differences between land use patterns in the more developed countries of the temperate zones and in the less developed countries of the tropics.

Map 7 World Population Density

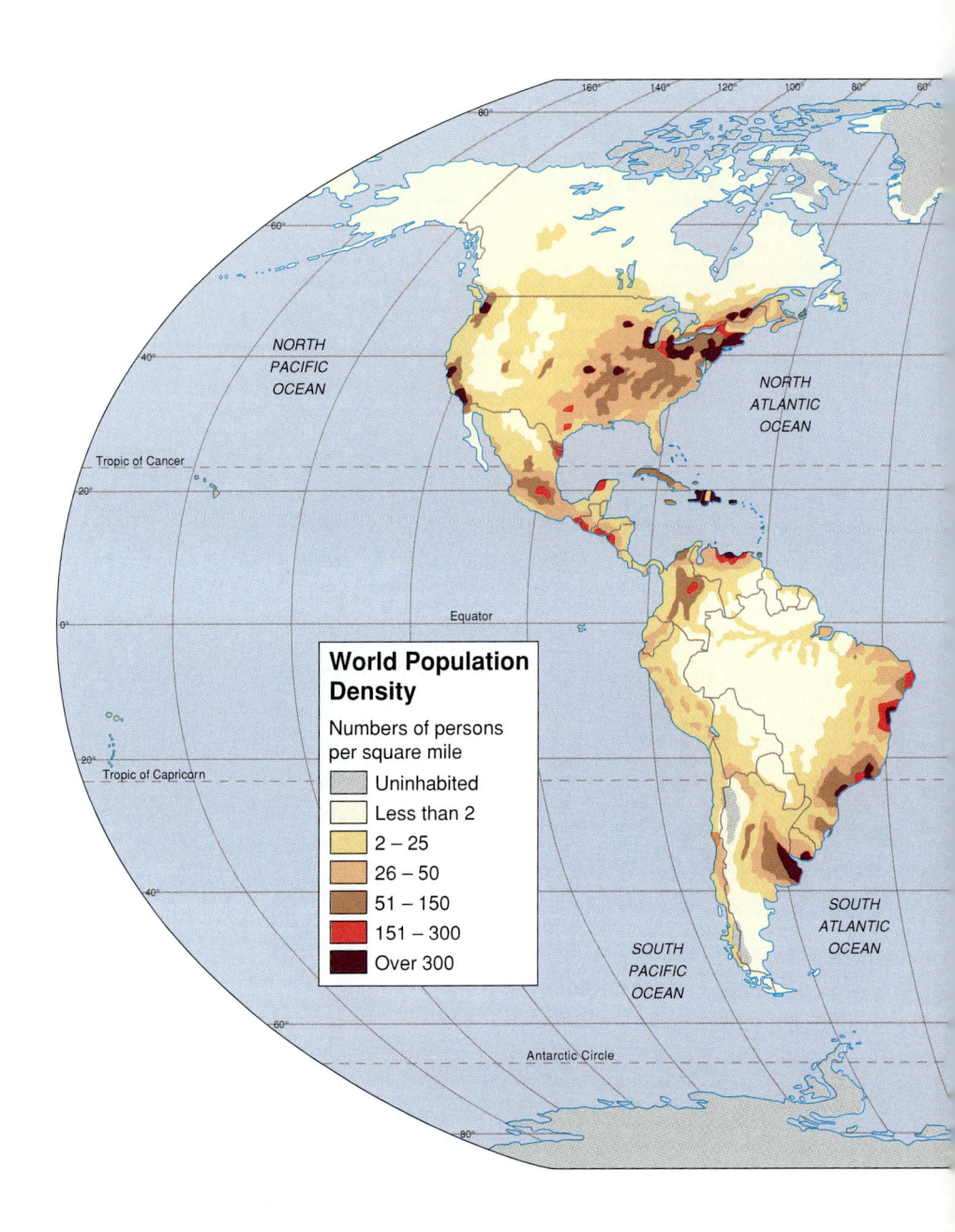

World Population Density

Numbers of persons per square mile

- Uninhabited
- Less than 2
- 2 – 25
- 26 – 50
- 51 – 150
- 151 – 300
- Over 300

No feature of human activity is more reflective of environmental conditions than where people live. In the areas of densest population, a mixture of natural and human factors have combined to allow maximum food production, maximum urbanization, and especially concentrated economic activity. Three such great concentrations of human population appear on the map—East Asia, South Asia, and Europe—with a fourth lesser concentration in eastern North America (the "Megalopolis" region of the United States and Canada). One of these great population clusters—South Asia—is still growing rapidly and can be expected to become even more densely populated by the beginning of the twenty-first century. The other concentrations are likely to remain about as they are now. In Europe and North America, this is the result of economic development that has caused population growth to level off during the last century. In East Asia, population has also begun to grow more slowly. In the case of Japan and the Koreas, this is the consequence of economic development; in the case of China, it is the consequence of government intervention in the form of strict family planning. The areas of future high density (in addition to those already existing) are likely to be in Latin America and Africa, where population growth rates are well above the world average. Population that is too large or growing at an excessive rate when measured against a region's habitability is one of the greatest inhibitors of economic development.

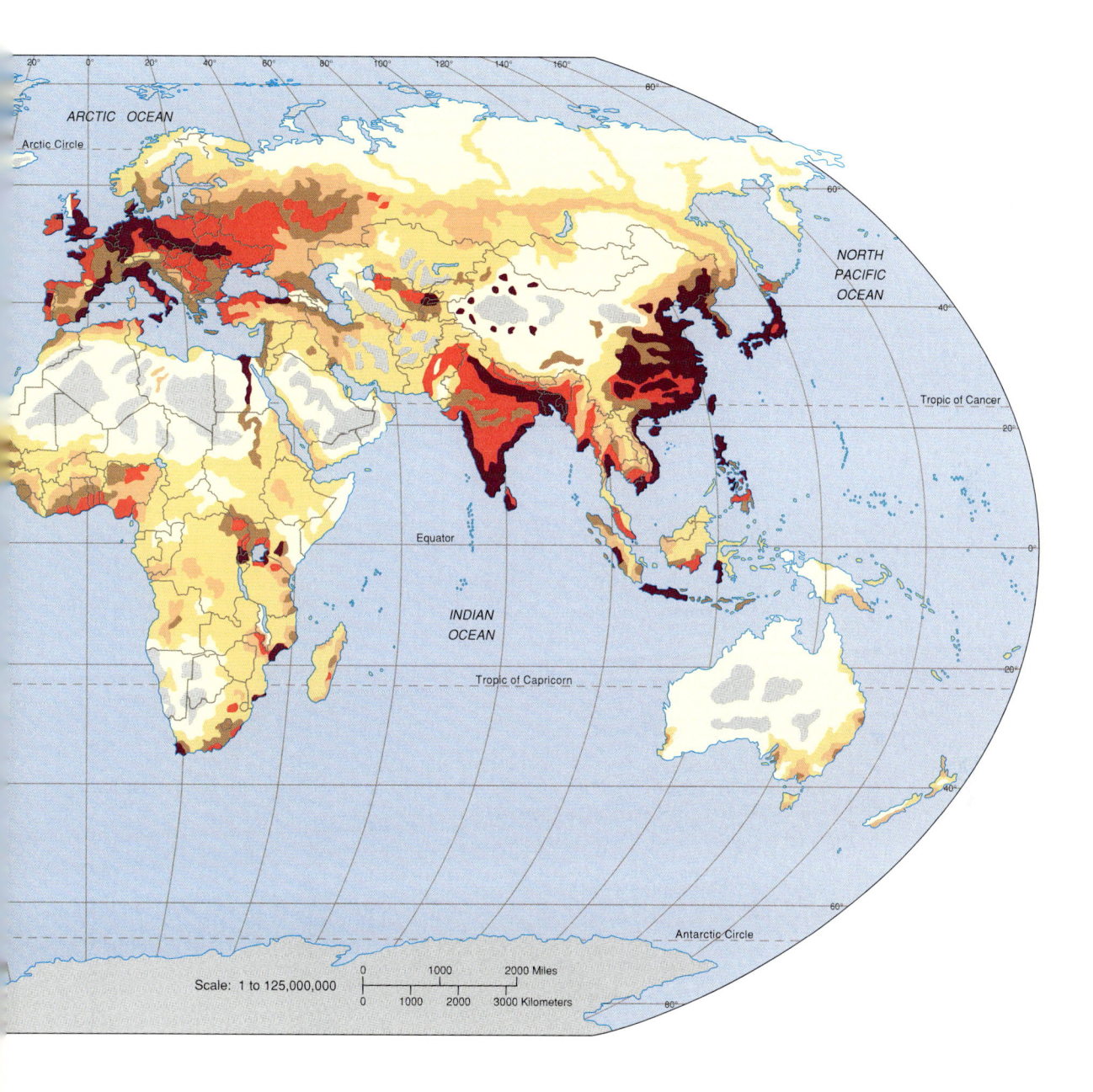

Table A
World Countries: Area, Population, and Population Density, 1996

COUNTRY	AREA (mi^2)	POPULATION (1996 est.)[a]	DENSITY (pop/mi^2)[b]
Afghanistan	251,826	21,252,000	84
Albania	11,100	3,414,000	308
Algeria	919,595	28,539,000	31
Andorra	175	66,000	377
Angola	481,354	10,070,000	21
Antigua and Barbuda	171	65,000	380
Argentina	1,073,400	34,293,000	32
Armenia	11,506	3,557,000	309
Australia	2,966,155	18,322,000	6
Austria	32,377	7,987,000	247
Azerbaijan	33,436	7,790,000	233
Bahamas	5,382	257,000	48
Bahrain	267	576,000	2,157
Bangladesh	55,598	128,095,000	2,304
Barbados	166	256,000	1,542
Belarus	80,155	10,437,000	130
Belgium	11,783	10,082,000	856
Belize	8,866	214,000	24
Benin	43,475	5,523,000	130
Bhutan	18,200	1,781,000	98
Bolivia	424,165	7,896,000	19
Bosnia-Herzegovina	19,741	3,202,000	162
Botswana	231,800	1,392,000	6
Brazil	3,286,488	160,737,000	49
Brunei	2,226	292,000	131
Bulgaria	42,823	8,755,000	205
Burkina Faso	105,869	10,423,000	98
Burundi	10,745	6,262,000	582
Cambodia	69,898	10,561,000	151
Cameroon	183,569	13,521,000	74
Canada	3,849,674	28,435,000	7
Cape Verde	1,557	436,000	280
Central African Republic	240,535	3,210,000	14
Chad	495,755	5,587,000	11
Chile	292,135	14,161,000	48
China	3,689,631	1,203,097,000	326
Colombia	440,831	35,200,000	80
Comoros	863	549,000	636
Congo	132,047	2,505,000	19
Costa Rica	19,730	3,419,000	173
Croatia	21,829	4,666,000	214
Cuba	42,804	10,938,000	255
Cyprus	3,593	737,000	205
Czech Republic	30,613	10,433,000	341
Denmark	16,638	5,199,000	312
Djibouti	8,958	510,000	57
Dominica	305	83,000	272
Dominican Republic	18,704	7,948,000	425
Ecuador	109,484	10,891,000	99
Egypt	386,662	62,360,000	161
El Salvador	8,124	5,870,000	723
Equatorial Guinea	10,831	420,000	39
Eritrea	45,300	3,579,000	79
Estonia	17,413	1,625,000	93
Ethiopia	483,123	55,979,000	116
Fiji	7,078	773,000	109
Finland	130,559	5,085,000	39
France	211,208	58,109,000	275
Gabon	103,347	1,156,000	11
Gambia	4,127	989,000	239
Georgia	26,911	5,726,000	213
Germany	137,882	81,338,000	590
Ghana	92,098	17,763,000	193
Greece	50,962	10,648,000	209

(Continued on next page)

COUNTRY	AREA (mi^2)	POPULATION (1996 est.)[a]	DENSITY (pop/mi^2)[b]
Grenada	133	94,000	707
Guatemala	42,042	10,999,000	262
Guinea	94,926	6,549,000	69
Guinea-Bissau	13,948	1,125,000	81
Guyana	83,000	724,000	8
Haiti	10,714	6,518,000	608
Honduras	43,277	5,460,000	126
Hungary	35,920	10,319,000	287
Iceland	36,769	266,000	7
India	1,237,062	936,546,000	756
Indonesia	742,410	203,584,000	274
Iran	632,457	64,625,000	102
Iraq	169,235	20,644,000	122
Ireland	27,137	3,550,000	131
Israel[c]	8,019	5,143,000	641
Italy	116,234	58,262,000	501
Ivory Coast	124,518	14,791,000	118
Jamaica	4,244	2,574,000	606
Japan	145,870	125,506,000	860
Jordan	35,135	4,101,000	117
Kazakhstan	1,049,156	17,377,000	17
Kenya	224,961	28,817,000	128
Kiribati	280	79,000	282
Korea, North	46,540	23,487,000	505
Korea, South	38,230	45,554,000	1,191
Kuwait	6,880	1,817,000	264
Kyrgyzstan	76,641	4,770,000	62
Laos	91,429	4,837,000	53
Latvia	24,595	2,763,000	112
Lebanon	4,015	3,696,000	921
Lesotho	11,720	1,993,000	170
Liberia	38,250	3,073,000	80
Libya	679,362	5,248,000	8
Liechtenstein	62	31,000	500
Lithuania	25,174	3,876,000	154
Luxembourg	998	405,000	406
Macedonia	9,928	2,160,000	218
Madagascar	226,658	13,862,000	61
Malawi	45,747	9,808,000	214
Malaysia	129,251	19,724,000	152
Maldives	115	261,000	2,250
Mali	478,767	9,375,000	20
Malta	122	370,000	3,032
Marshall Islands	70	56,000	802
Mauritania	395,956	2,263,000	6
Mauritius	788	1,127,000	1,430
Mexico	756,066	93,986,000	124
Micronesia	271	123,000	454
Moldova	13,012	4,490,000	345
Mongolia	604,829	2,494,000	4
Morocco[d]	275,114	28,169,000	102
Mozambique	308,642	18,150,000	59
Myanmar	261,228	45,104,000	173
Namibia	317,818	1,652,000	5
Nepal	56,827	21,561,000	386
Netherlands	16,133	15,543,000	963
New Zealand	103,519	3,407,000	33
Nicaragua	50,054	4,206,000	84
Niger	489,191	9,280,000	19
Nigeria	356,669	101,232,000	284
Norway	125,050	4,331,000	35
Oman	82,030	2,125,000	26
Pakistan	310,432	131,542,000	424
Palau	188	17,000	88
Panama	29,157	2,681,000	92
Papua New Guinea	178,704	4,295,000	24
Paraguay	157,048	5,358,000	34
Peru	496,225	24,087,000	49
Philippines	115,831	73,087,000	633

(Continued on next page)

COUNTRY	AREA (mi^2)	POPULATION (1996 est.)[a]	DENSITY (pop/mi^2)[b]
Poland	120,728	38,792,000	321
Portugal	35,516	10,562,000	297
Qatar	4,416	534,000	121
Romania	91,699	23,198,000	253
Russia	6,592,849	149,909,000	23
Rwanda	10,169	8,605,000	846
St. Kitts and Nevis	104	41,000	394
St. Lucia	238	156,000	655
St. Vincent/Grenadines	150	118,000	787
San Marino	24	24,000	1,013
São Tomé and Príncipe	372	140,000	376
Saudi Arabia	830,000	18,730,000	23
Senegal	75,951	9,907,000	130
Serbia-Montenegro (Yugoslavia)	39,449	11,102,000	281
Seychelles	175	73,000	417
Sierra Leone	27,925	4,753,000	170
Singapore	246	2,889,000	11,744
Slovakia	18,933	5,432,000	287
Slovenia	7,819	2,052,000	262
Solomon Islands	10,954	399,000	36
Somalia	246,201	7,348,000	30
South Africa	433,680	45,095,000	104
Spain	194,885	39,404,000	202
Sri Lanka	24,962	18,343,000	735
Sudan	967,500	30,120,000	31
Suriname	63,251	430,000	7
Swaziland	6,704	967,000	144
Sweden	173,732	8,822,000	51
Switzerland	15,943	7,085,000	444
Syria	71,498	15,452,000	216
Taiwan	13,900	21,501,000	1,547
Tajikistan	55,251	6,155,000	111
Tanzania	364,900	28,701,000	79
Thailand	198,115	60,271,000	304
Togo	21,925	4,410,000	201
Tonga	290	106,000	366
Trinidad and Tobago	1,980	1,271,000	642
Tunisia	63,170	8,880,000	141
Turkey	300,948	63,406,000	211
Turkmenistan	188,456	4,075,000	22
Tuvalu	9	10,000	1,063
Uganda	93,104	19,573,000	210
Ukraine	233,090	51,868,000	223
United Arab Emirates	32,278	2,925,000	88
United Kingdom	94,248	58,295,000	618
United States	3,787,425	263,814,000	70
Uruguay	68,500	3,223,000	47
Uzbekistan	172,742	23,089,000	134
Vanuatu	4,707	174,000	37
Venezuela	352,145	21,005,000	60
Vietnam	128,066	74,393,000	581
Western Samoa	1,093	209,000	191
Yemen	205,356	14,728,000	72
Zaire	905,446	44,061,000	49
Zambia	290,586	9,446,000	33
Zimbabwe	150,873	11,140,000	74

[a] All populations are estimates for mid-1996, rounded to the nearest 1,000 of the population.

[b] Population densities as calculated by the UN. The figures are close to but not the same as may be derived by dividing the estimated 1996 population of a state by its area, as shown in this table.

[c] The figures for Israel do not include the West Bank and Gaza. The former has an area of approximately 2,300 square miles, a population of 1.7 million, and a density of 746 persons per square mile; the latter has an area of 100 square miles, a population of nearly 800,000 persons, and an estimated population density of 7,600 persons per square mile.

[d] Figures for Morocco include Western Sahara, annexed by Morocco.

Sources: U.S. Bureau of the Census, World Population Profile; United Nations Population Division; Information Please Almanac 1996; The World Almanac and Book of Facts 1996.

Part II

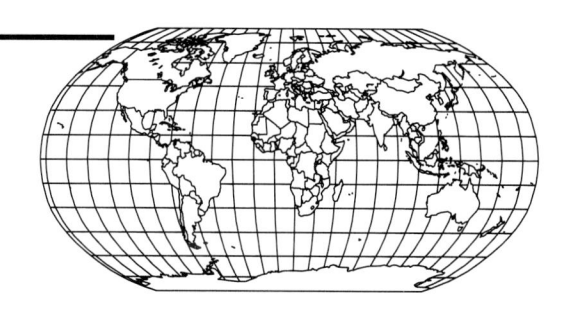

The Human Factor: The World's People

Map 8 Population Growth Rates
Map 9 Total Fertility Rate
Map 10 Infant Mortality Rate
Map 11 Average Life Expectancy at Birth
Map 12 Population by Age Group
Map 13 World Daily Per Capita Food Supply
Map 14 Child Malnutrition
Map 15 Illiteracy Rate
Map 16 Primary School Enrollment
Map 17 Total Labor Force
Map 18 Female Labor Force
Map 19 Female/Male Inequality in Education and Employment
Map 20 Urban Population
Map 21 Countries with Significant Minority Group Populations
Map 22 Quality of Life: The Human Development Index
Table B Size and Growth of Population, 1950–2025
Table C World Countries: Demography, 1975–1995
Table D Mortality, Health, and Nutrition, 1970–1995
Table E Education and Literacy, 1970–1993

Map 8 Population Growth Rates

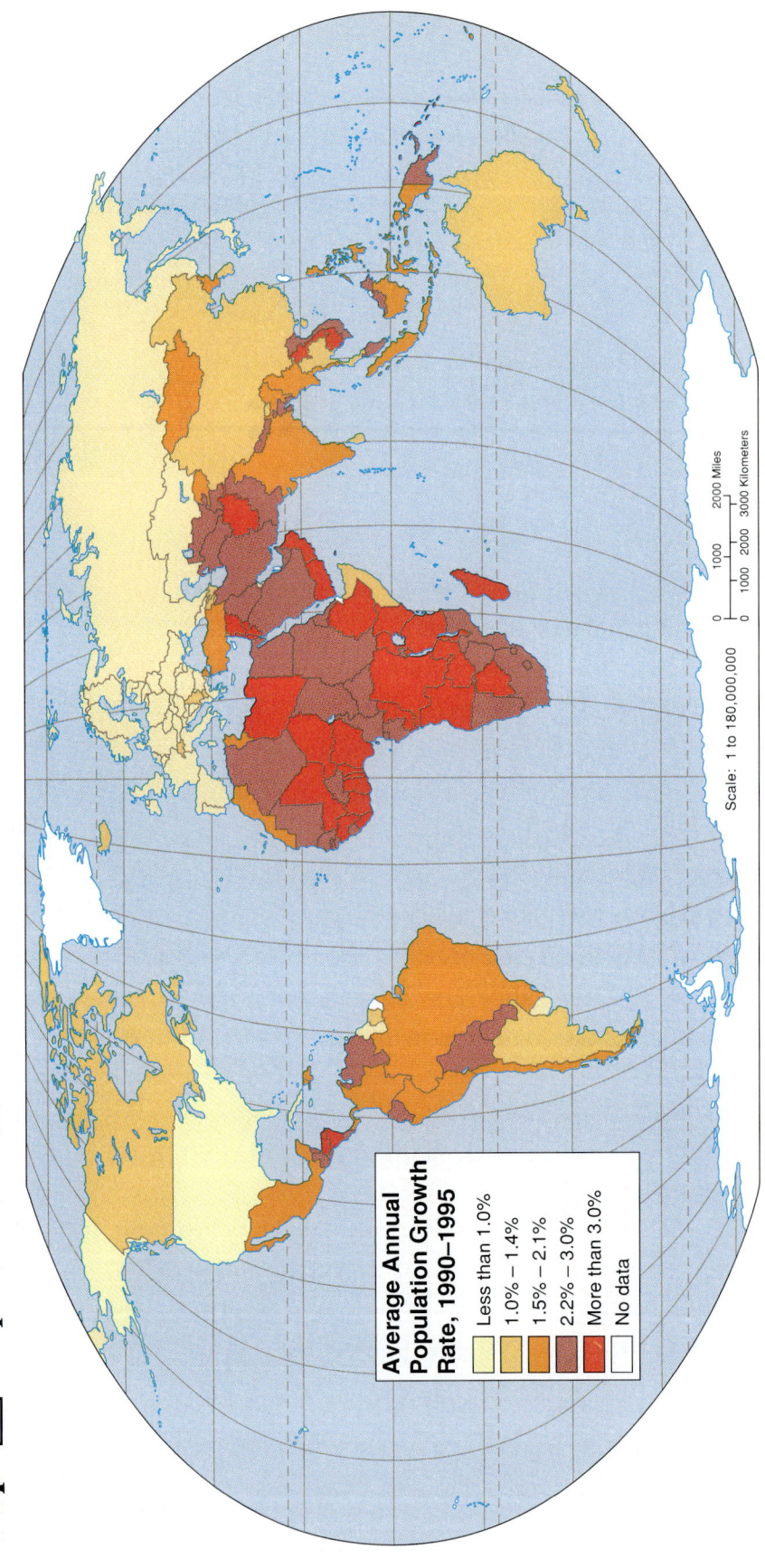

Average Annual Population Growth Rate, 1990–1995
- Less than 1.0%
- 1.0% – 1.4%
- 1.5% – 2.1%
- 2.2% – 3.0%
- More than 3.0%
- No data

Scale: 1 to 180,000,000

Of all the statistical measurements of human population, that of the rate of population growth is the most important. The growth rate of a population is a combination of natural change (births and deaths), in-migration, and out-migration; it is obtained by adding the number of births to the number of immigrants during a year and subtracting from that total the sum of deaths and emigrants for the same year. For a specific country, this figure will determine many things about the country's future ability to feed, house, educate, and provide medical services to its citizens. Some of the countries with the largest populations (such as India) also have high growth rates. Since these countries tend to be in developing regions, the combination of high population and high growth rates poses special problems for continuing economic development and carries heightened risks of environmental degradation. Many people believe that the rapidly expanding world population is a potential crisis that may cause environmental and human disaster by the middle of the twenty-first century.

— 20 —

Map 9 Total Fertility Rate

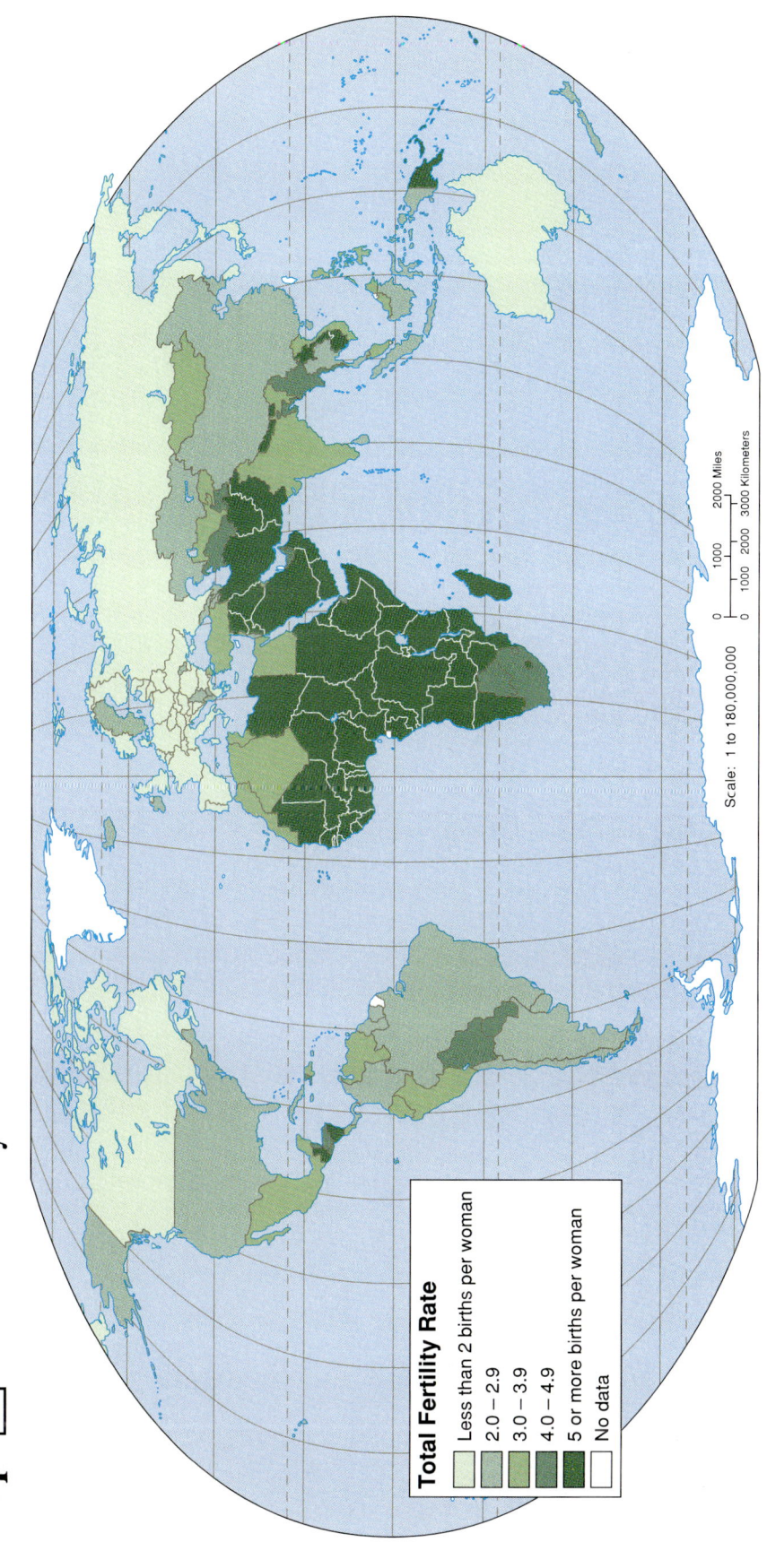

Total Fertility Rate
- Less than 2 births per woman
- 2.0 – 2.9
- 3.0 – 3.9
- 4.0 – 4.9
- 5 or more births per woman
- No data

Scale: 1 to 180,000,000

The fertility rate measures the number of children that a woman is expected to bear during her lifetime, based on the age-specific fertility figures of women between 15 and 40 (the normal childbearing years). While fertility rates tell us a great deal about present population growth, with high fertility rates indicating high population growth rates, they are also indicative of potential or projected growth. A country whose women can be expected to bear many children is a country with enormous potential for population growth in the future. Given present fertility rates, for example, the number of offspring from the average German woman over the next three generations (the total number of children, grandchildren, and great-grandchildren) will be 7. During the same three generations, the average American woman will have a total of 17 children, grandchildren, and great-grandchildren. But during this time, assuming that present fertility rates are maintained, the average woman in sub-Saharan Africa will have *258* children, grandchildren, and great-grandchildren. You might be interested in working out some potential population growth rates over two or three generations, using the data as presented on the map.

Note: Total fertility rate figures are based on yearly age-specific fertility rate data from 1990 to 1995.

— 21 —

Map 10 Infant Mortality Rate

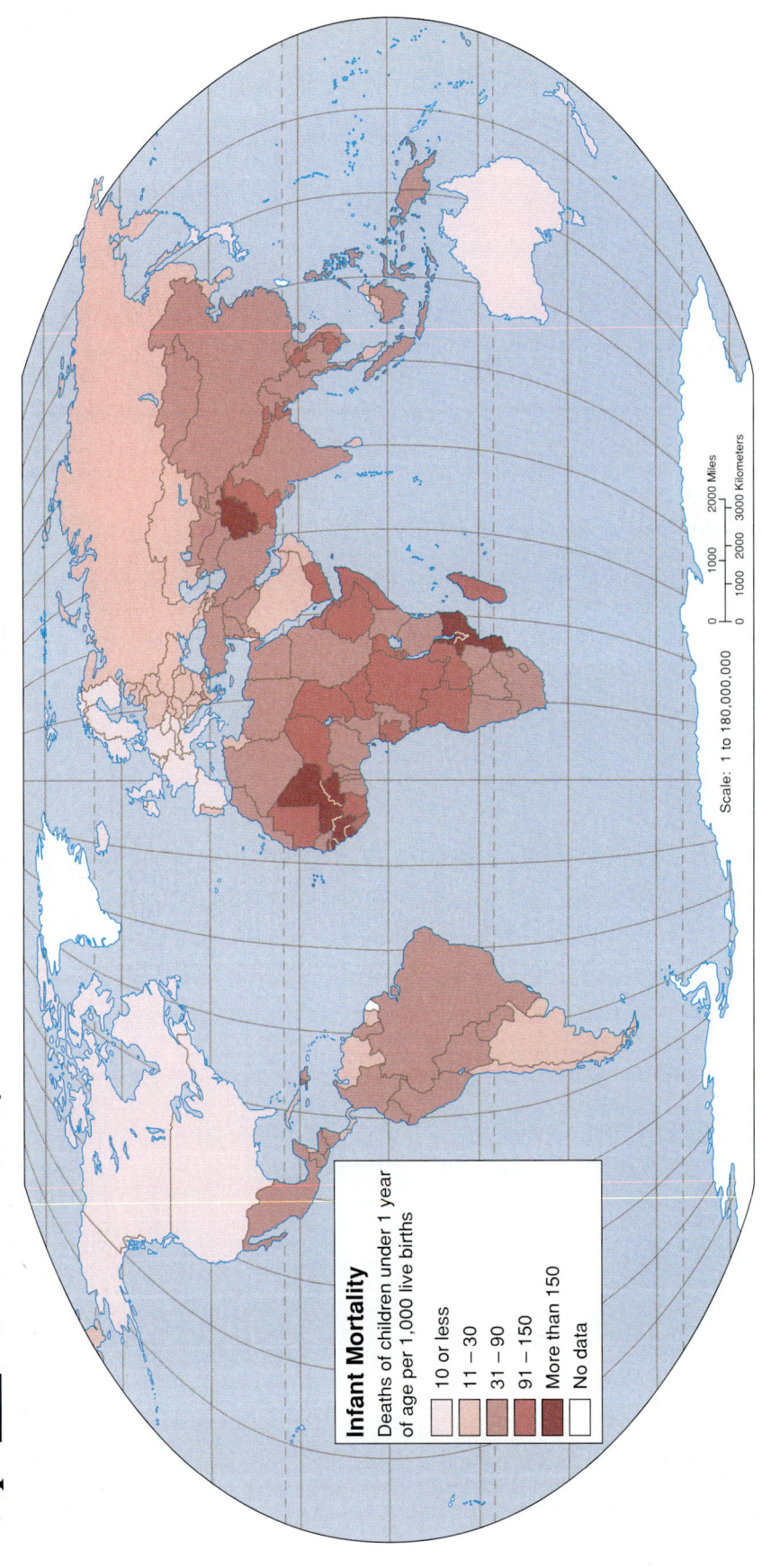

Infant Mortality
Deaths of children under 1 year of age per 1,000 live births

- 10 or less
- 11 – 30
- 31 – 90
- 91 – 150
- More than 150
- No data

Scale: 1 to 180,000,000

Infant mortality rates are calculated by dividing the number of children born in a given year who die before their first birthday by the total number of children born that year and then multiplying by 1,000; this shows how many infants have died for every 1,000 births. Infant mortality rates are prime indicators of economic development. In highly developed economies, with advanced medical technologies, sufficient diets, and adequate public sanitation, the infant mortality rates tend to be quite low. By contrast, in less developed countries, with the disadvantages of poor diet, limited access to medical technology, and the other problems of poverty, infant mortality rates tend to be high. Although worldwide infant mortality has decreased significantly during the last 2 decades, many regions of the world still experience infant mortality above the 10 percent level (100 deaths per 1,000 live births). Such infant mortality rates not only represent human tragedy at its most basic level, but also are powerful inhibiting factors on the future of human development. Comparing the differences in infant mortality rates between the North and the South shows that children in most African countries are more than 10 times as likely to die before their first birthday than children in some northern countries, such as Canada.

Map 11 Average Life Expectancy at Birth

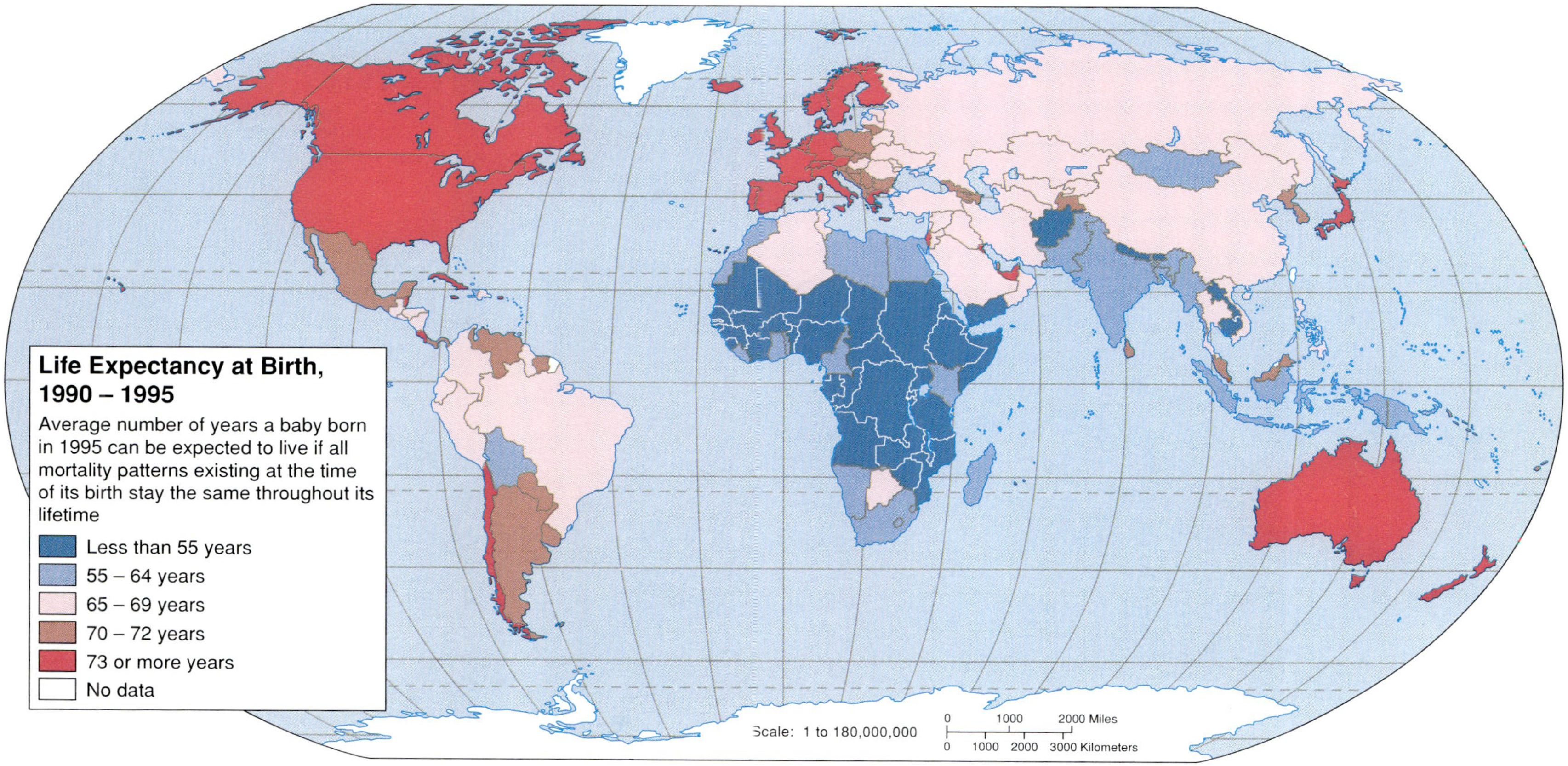
Life Expectancy at Birth, 1990 – 1995
Average number of years a baby born in 1995 can be expected to live if all mortality patterns existing at the time of its birth stay the same throughout its lifetime
Less than 55 years
55 – 64 years
65 – 69 years
70 – 72 years
73 or more years
No data
Scale: 1 to 180,000,000
0 1000 2000 Miles
0 1000 2000 3000 Kilometers

Average life expectancy at birth is a measure of the average longevity of the population of a country. Like all average measures, it is distorted by extremes. For example, a country with a high mortality rate among children will have a low average life expectancy. Thus, an average life expectancy of 45 years does not mean that everyone can be expected to die at the age of 45. More normally, what the figure means is that a substantial number of children die between birth and 5 years of age, thus reducing the average life expectancy for the entire population. In spite of the dangers inherent in misinterpreting the data, average life expectancy (along with infant mortality and several other measures) is a valid way of judging the relative health of a population. It reflects the nature of the health care system, public sanitation and disease control, nutrition, and a number of other key human need indicators. As such, it is measure of well-being that is significant in indicating economic development.

Map 12 Population by Age Group

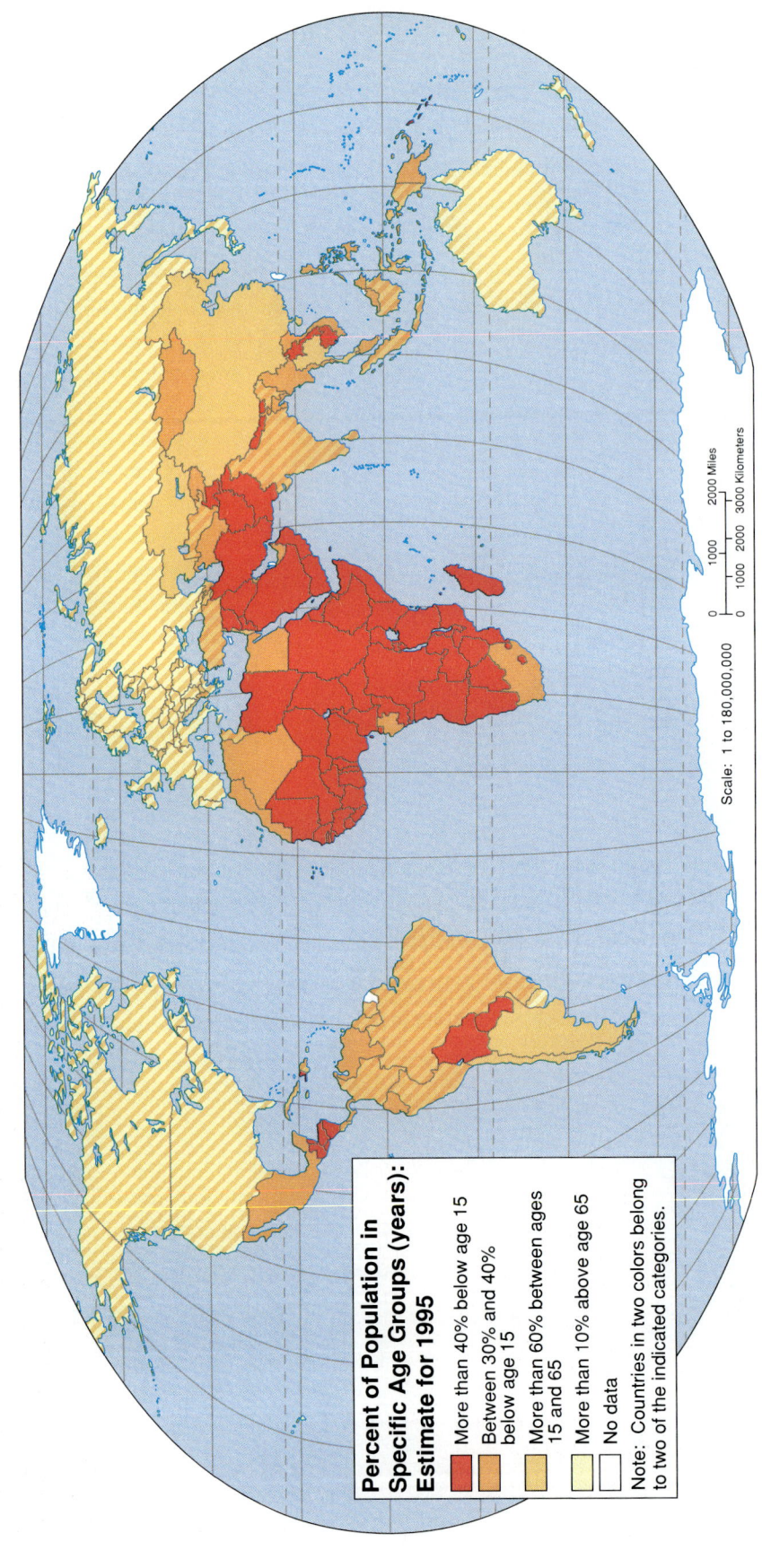

Percent of Population in Specific Age Groups (years): Estimate for 1995

- More than 40% below age 15
- Between 30% and 40% below age 15
- More than 60% between ages 15 and 65
- More than 10% above age 65
- No data

Note: Countries in two colors belong to two of the indicated categories.

Scale: 1 to 180,000,000

Of all the measurements that illustrate the dynamics of a population, age distribution may be the most significant, particularly when viewed in combination with average growth rates. The particular relevance of age distribution is that it tells us what to expect from a population in terms of growth over the next generation. If, for example, approximately 40–50 percent of a population is below the age of 15, that suggests that in the next generation about one-quarter of the total population will be women of childbearing age. When age distribution is combined with fertility rates (the average number of children born per woman in a population), an especially valid measurement of future growth potential may be derived. A simple example: Nigeria, with a 1990 population of 109 million, has 50 percent of its population below the age of 15 and a fertility rate of 6.6; the United States, with a 1990 population of 250 million, has 21 percent of its population below the age of 15 and a fertility rate of 1.9. During the period in which those persons presently under the age of 15 are in their childbearing years, Nigeria can be expected to add a total of approximately 180 million persons to its total population. Over the same period, the United States can be expected to add only 50 million. Such differences are important in estimating the potential of different areas for continuing economic development.

– 24 –

Map 13 World Daily Per Capita Food Supply

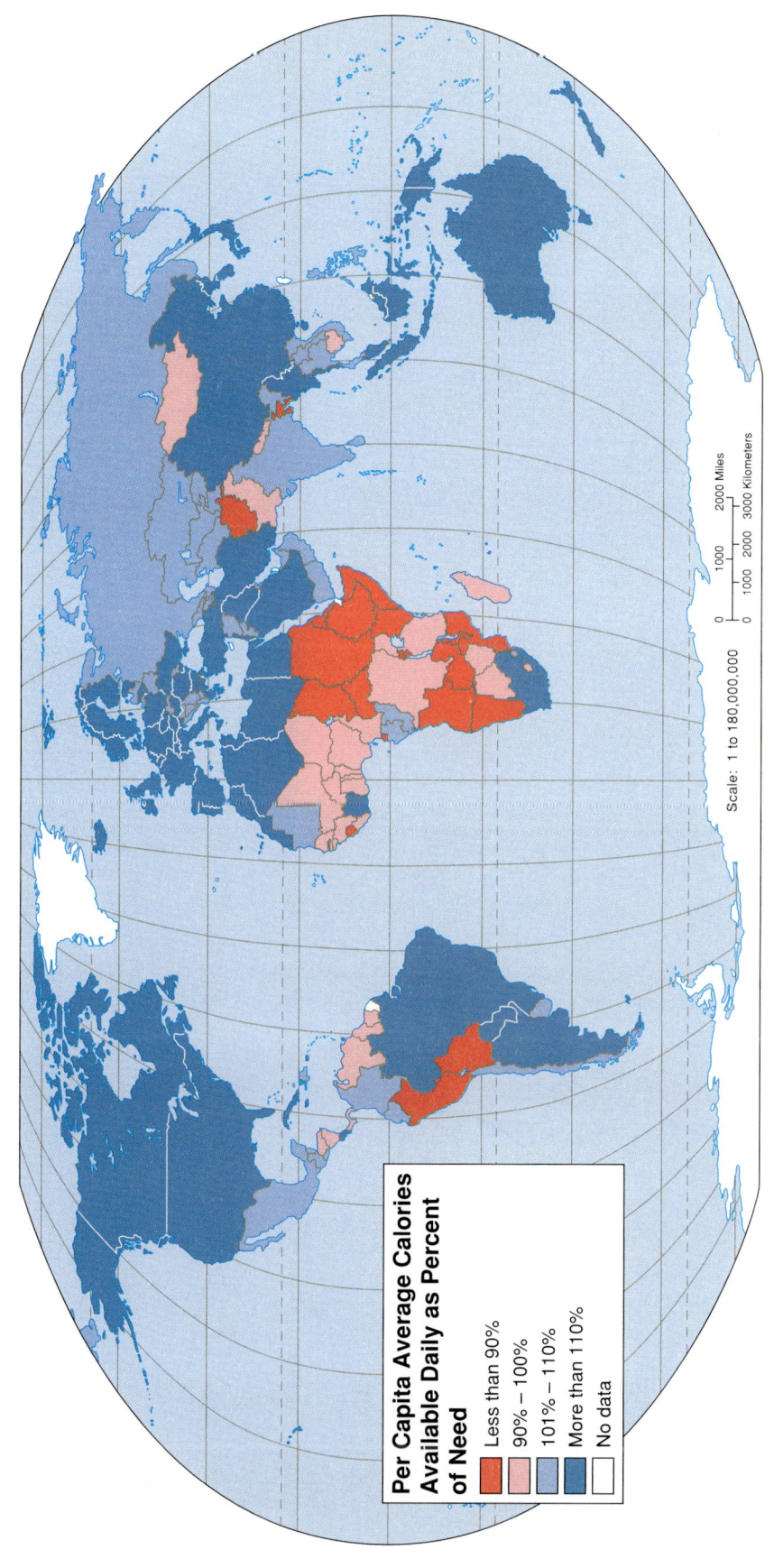

Per Capita Average Calories Available Daily as Percent of Need

- Less than 90%
- 90% – 100%
- 101% – 110%
- More than 110%
- No data

Scale: 1 to 180,000,000

0   1000   2000 Miles
0 1000 2000 3000 Kilometers

The data shown on this map indicate the presence or absence of critical food shortages. While they do not necessarily indicate the presence of starvation or famine, they certainly do indicate potential problem areas for the next decade. The measurements are in calories from *all* food sources: domestic production, international trade, drawdown on stocks or food reserves, and direct foreign contributions or aid. The quantity of calories available is that amount, estimated by the UN's Food and Agriculture Organization (FAO), that actually reaches consumers. The calories actually consumed may be lower than the figures shown, depending on how much is lost in a variety of ways: in home storage (to pests such as rats and mice), in preparation and cooking, through consumption by pets and domestic animals, and as discarded foods, for example. The estimate of need is not a uniform global value but is calculated for each country on the basis of the age and sex distribution of the population and the estimated level of activity of the population.

— 25 —

Map 14 Child Malnutrition

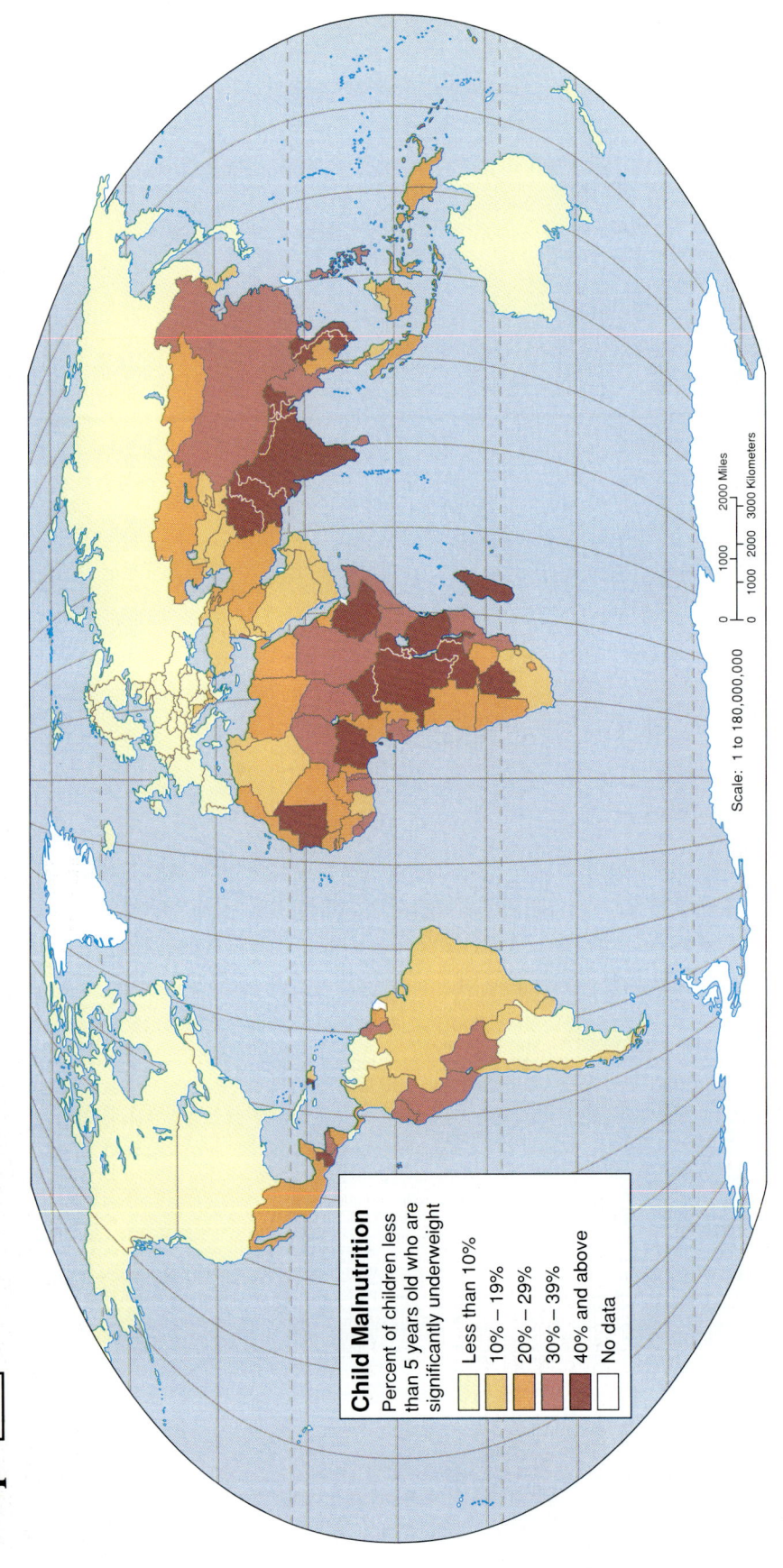

Child Malnutrition
Percent of children less than 5 years old who are significantly underweight

- Less than 10%
- 10% – 19%
- 20% – 29%
- 30% – 39%
- 40% and above
- No data

Scale: 1 to 180,000,000

The weight of poverty is not evenly spread among the members of a population, falling disproportionately upon the weakest and most disadvantaged members of society. In most societies, these individuals are children, particularly female children. Children simply do not compete as successfully as adults for their (meager) share of the daily food supply. Where food shortages prevail, children tend to have the future quality of their lives severely compromised by poor nutrition, which, in a downward spiral, robs them of the energy necessary to compete more effectively for food. Children who are inadequately fed are less likely to do well in school, are more prone to debilitating disease, and will more often become a drain on scarce societal resources than well-fed children.

– 26 –

Map 15 Illiteracy Rate

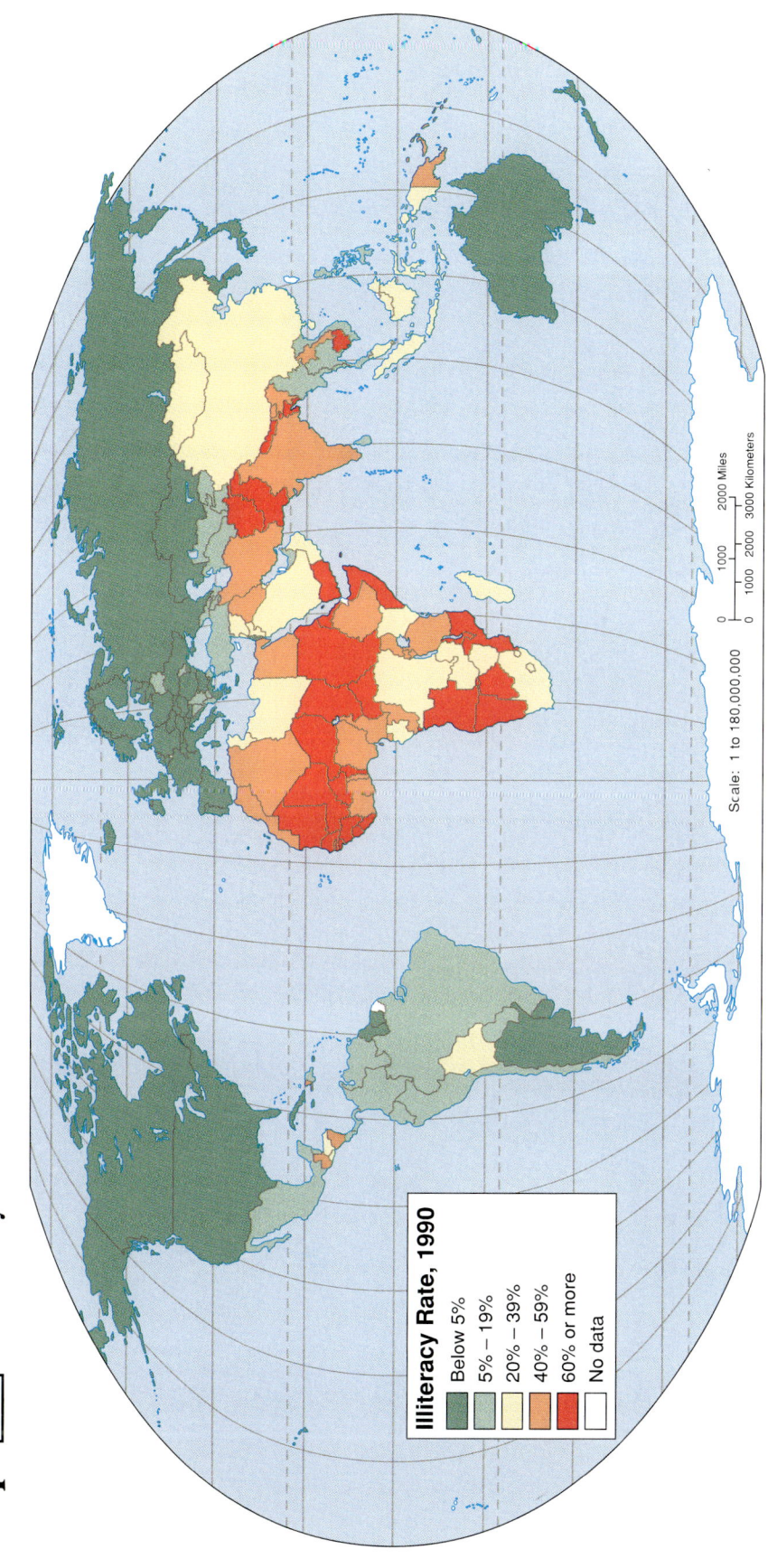

Illiteracy Rate, 1990
- Below 5%
- 5% – 19%
- 20% – 39%
- 40% – 59%
- 60% or more
- No data

The gains in living standards that developing countries have experienced during the last 2 decades are manifested in two major areas: life expectancy and literacy. The increase in global literacy is largely the consequence of an increase in primary school enrollment, particularly throughout Middle and South America, Africa, and Asia. Worldwide, education is perceived as a way to advance economic status. Unfortunately, although gains have been made, there are still countries where illiteracy rates—particularly among the females of the population—are well above global norms. The long-term potential of these countries is severely compromised as a result.

— 27 —

Map 16 Primary School Enrollment

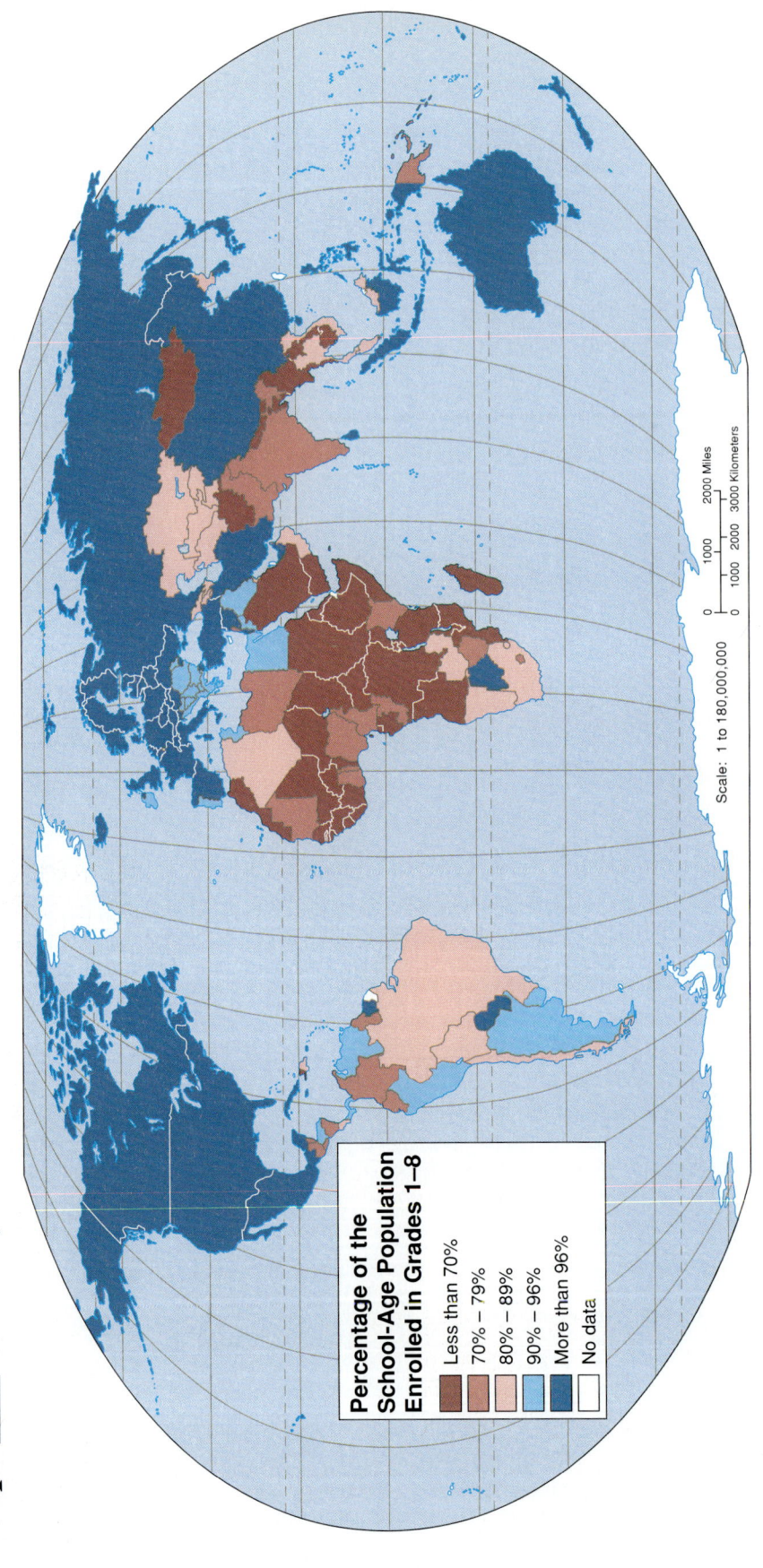

Percentage of the School-Age Population Enrolled in Grades 1–8
- Less than 70%
- 70% – 79%
- 80% – 89%
- 90% – 96%
- More than 96%
- No data

Scale: 1 to 180,000,000

Like many of the other measures illustrated in this atlas, primary school enrollment is a clear reflection of the division of the world into "have" and "have-not" countries. It is also a measure that has changed more rapidly over the last decade than demographic and other indicators of development, as countries of even very modest means have made concerted attempts to attain relatively high percentages of primary school enrollment. That they have been able to do so is good evidence of the fact that reasonably respectable levels of human development are feasible at even modest income levels. High primary school enrollment is also a reflection of the worldwide opinion that a major element in economic development is a well-educated, literate population. The links between human progress, as typified by higher levels of education, and economic growth are not automatic, however, and those countries without programs for maintaining the headway gained by improved education may be on the road to failure in terms of economic development.

Map 17 Total Labor Force

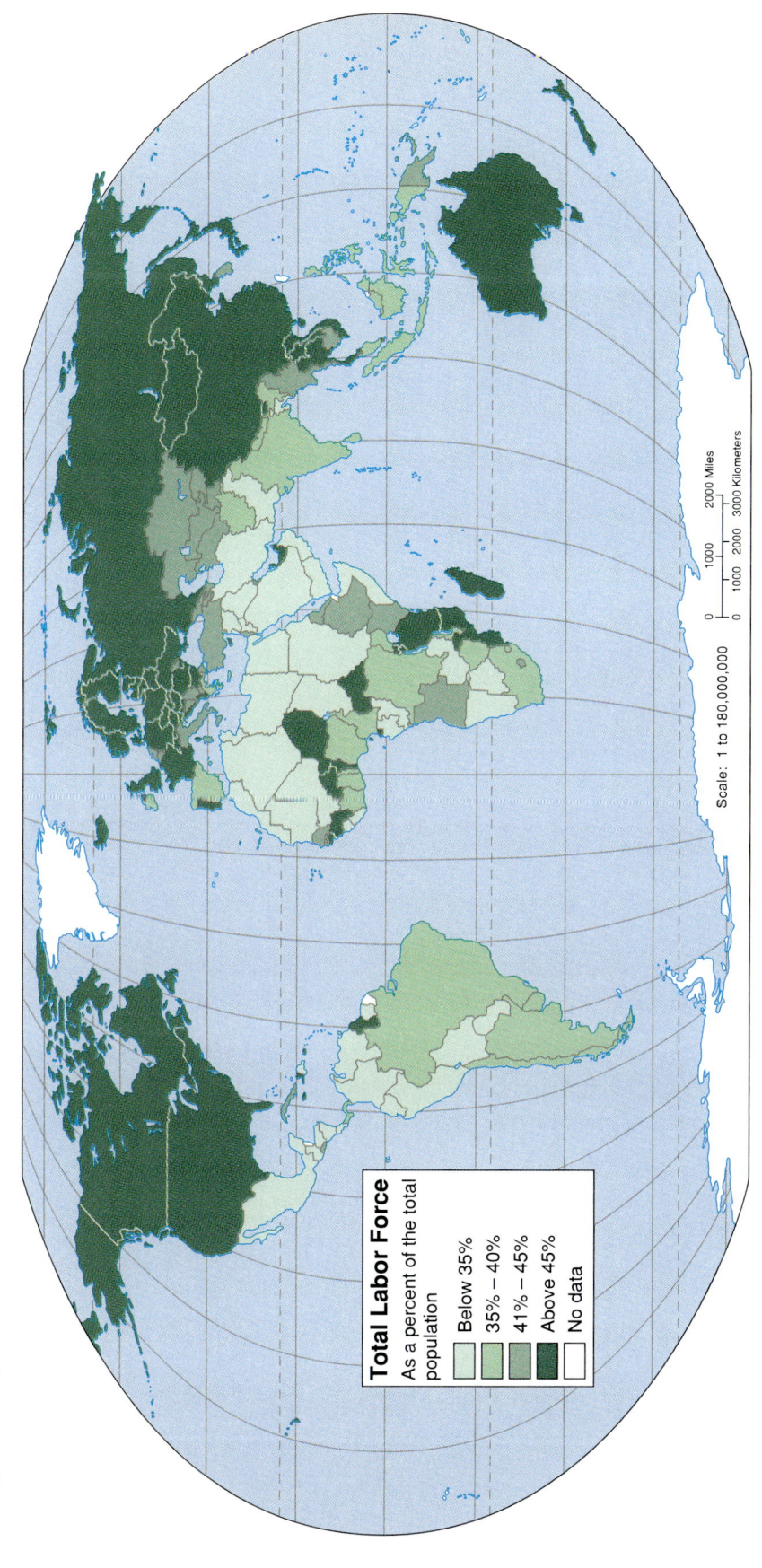

Total Labor Force
As a percent of the total population

- Below 35%
- 35% – 40%
- 41% – 45%
- Above 45%
- No data

Scale: 1 to 180,000,000

The term *labor force* refers to the economically active portion of a population, that is, all people who work or are without work but are available for and are seeking work to produce economic goods and services. The total labor force thus includes both the employed and the unemployed (as long as they are actively seeking employment). Labor force is considered a better indicator of economic potential than employment/unemployment figures, since unemployment figures will contain experienced workers with considerable potential who are temporarily out of work. Unemployment figures will also include persons seeking employment for the first time. Generally, countries with higher percentages of their total population within the labor force will be countries with higher levels of economic development. This is partly a function of levels of education and training and partly a function of the age distribution of populations. In developing countries, substantial percentages of the total population are too young to be part of the labor force.

– 29 –

Map 18 Female Labor Force

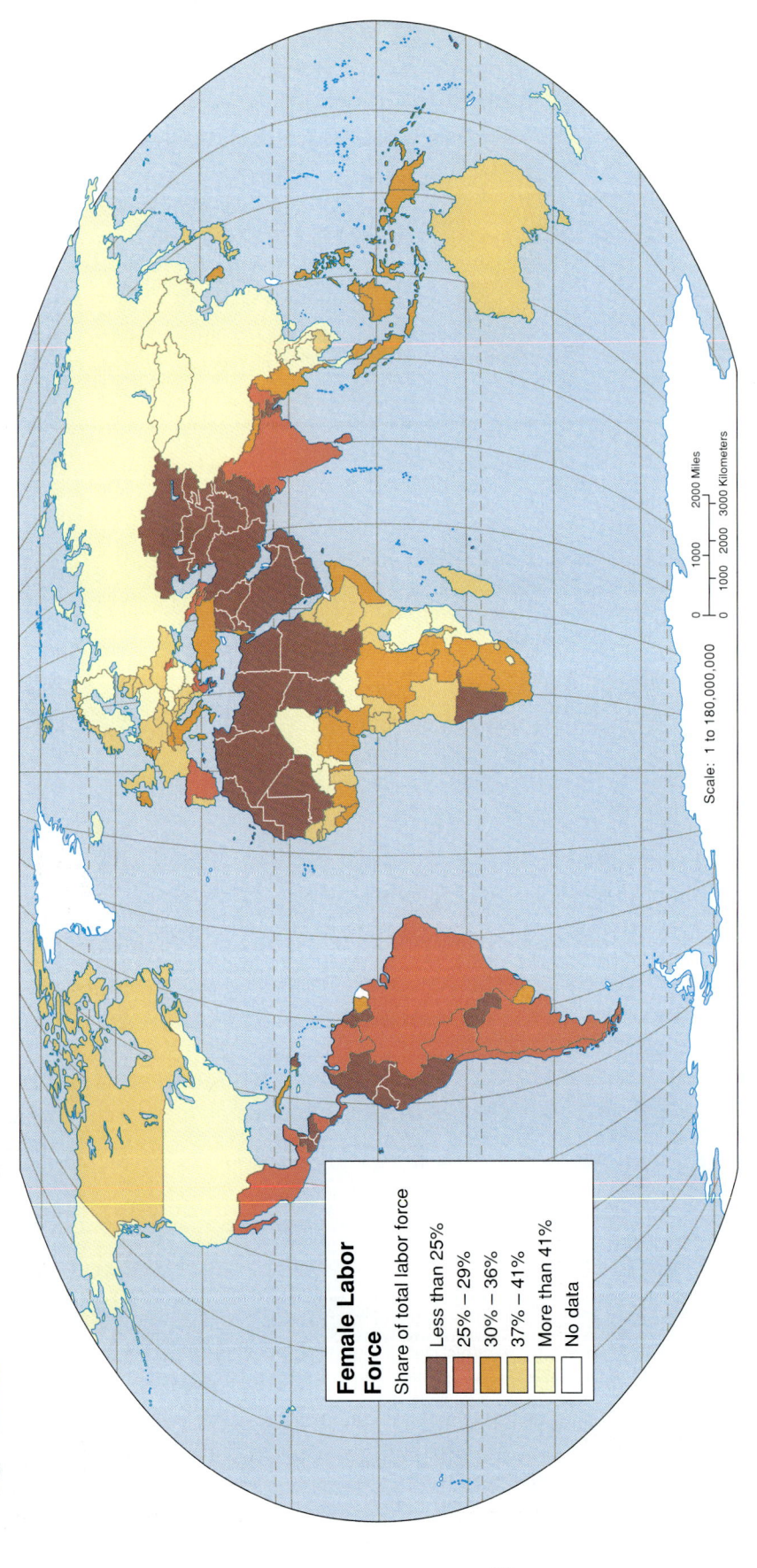

Female Labor Force
Share of total labor force
- Less than 25%
- 25% – 29%
- 30% – 36%
- 37% – 41%
- More than 41%
- No data

The percentage of the labor force that is composed of females is an intriguing measure of economic development. In general, countries with higher levels of economic development have larger percentages of women in the "formal" labor force (which excludes household and subsistence agricultural workers, among others). But a considerable number of countries with low percentages of women in the labor force rank somewhere in the middle on the economic development scale. There are two primary reasons why fewer females work in the formal sectors in these countries. One of the reasons is cultural bias. For example, in Latino regions of Middle and South America and in Islamic countries of North Africa and Southwest Asia, cultural bias against women tends to keep them out of the workplace and in the home, where they engage in nonreported household activities. A second reason for the absence of women from the labor force has to do with the structure of national or regional economies. In Native American regions of Middle and South America, in parts of Africa, and in parts of South and Southeast Asia, the role of women in subsistence agriculture is a significant one but is not reported as "employment," just as household activities are not reported as employment. Ultimately, the real significance of the percentage of females in the labor force may be as an indicator of a country's *potential* development and room for growth in those nonhousehold and nonsubsistence economic activities in which women are currently underemployed.

– 30 –

Map 19 Female/Male Inequality in Education and Employment

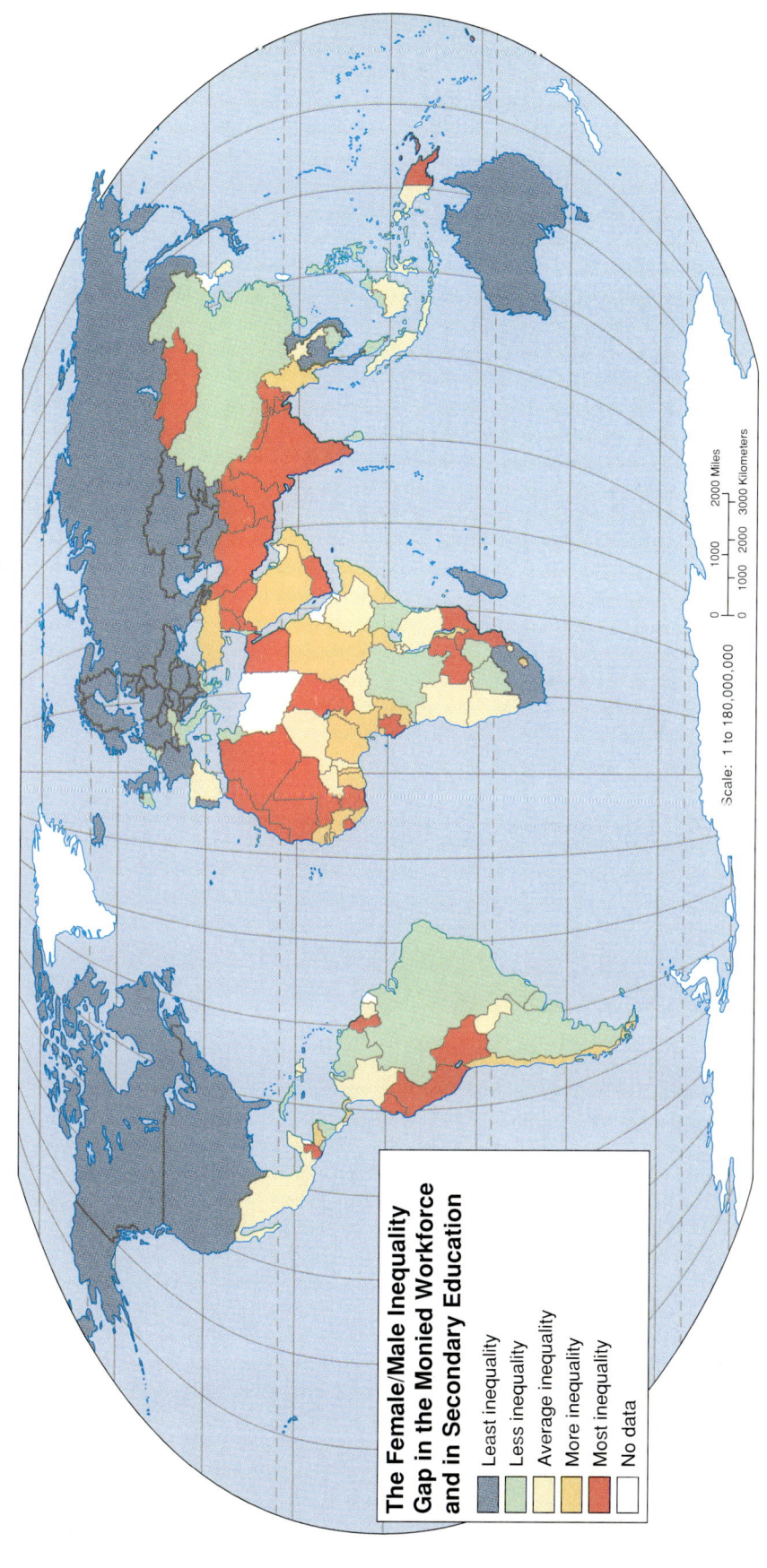

The Female/Male Inequality Gap in the Monied Workforce and in Secondary Education
- Least inequality
- Less inequality
- Average inequality
- More inequality
- Most inequality
- No data

Scale: 1 to 180,000,000

While women in developed countries, particularly in North America and Europe, have made significant advances in socioeconomic status in recent years, in most of the world females suffer from significant inequality when compared with their male counterparts. Although women have received the right to vote in most of the world's countries, in over 90 percent of these countries that right has only been granted to women in the last 50 years. In most regions, literacy rates for women still fall far short of those for men; in Africa and Asia, for example, only about half as many women are literate as men. Women marry considerably younger than men and attend school for shorter periods of time. The two criteria shown on this map are perhaps the most telling indicators of the unequal status of women in most of the world. Lack of secondary education in comparison with men prevents women from entering the workforce with equally high-paying jobs. Even where women are employed in positions similar to those held by men, they still tend to receive less compensation. The gap between rich and poor involves not only a clear geographic differentiation but a clear gender differentiation as well.

– 31 –

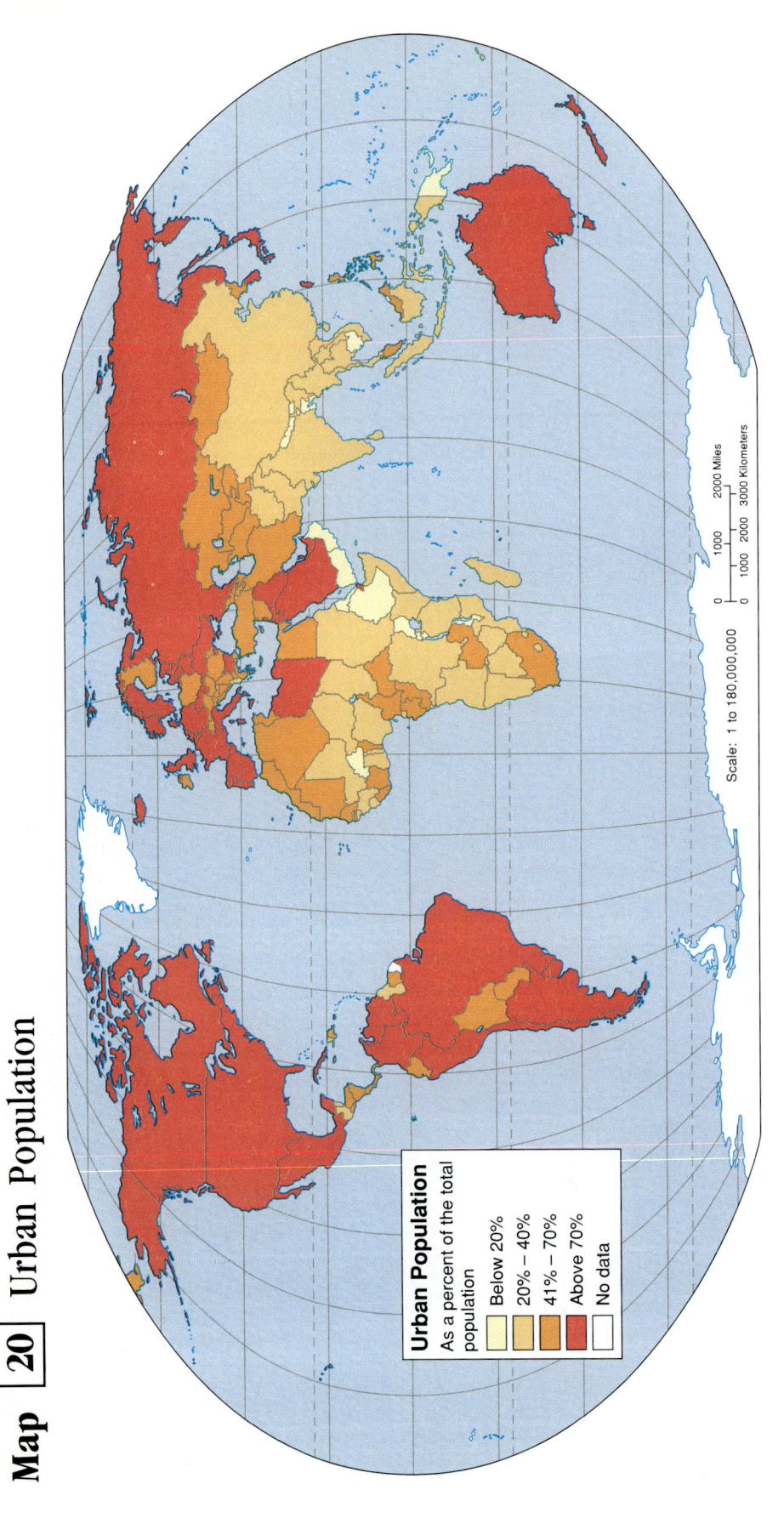

Map 20 Urban Population

Urban Population
As a percent of the total population

- Below 20%
- 20% – 40%
- 41% – 70%
- Above 70%
- No data

Scale: 1 to 180,000,000

The proportion of a country's population that resides in urban areas was formerly considered to be a measure of relative economic development, with countries possessing a large urban population ranking high on the development scale and countries with a more rural population ranking low. Given the rapid rate of urbanization in developing countries, however, this traditional measure no longer is as valuable. What relative urbanization rates now tell us is something about levels of economic development in a negative sense. Latin American, African, and Asian countries with more than 40 percent of their populations living in urban areas are quite probably suffering from a variety of problems: rural overpopulation and flight from the land, urban poverty and despair, high unemployment, and poor public services. The rate of urbanization in less developed nations is such that many cities in these nations will outstrip those in North America and Europe by the end of this century. It has been estimated, for example, that Mexico City—now the world's largest metropolis—may have as many as 27 million inhabitants by the year 2000. Urbanization was once viewed as an indicator of economic health. For many countries it is now a picture of potential economic and environmental disaster.

– 32 –

Map 21 Countries with Significant Minority Group Populations

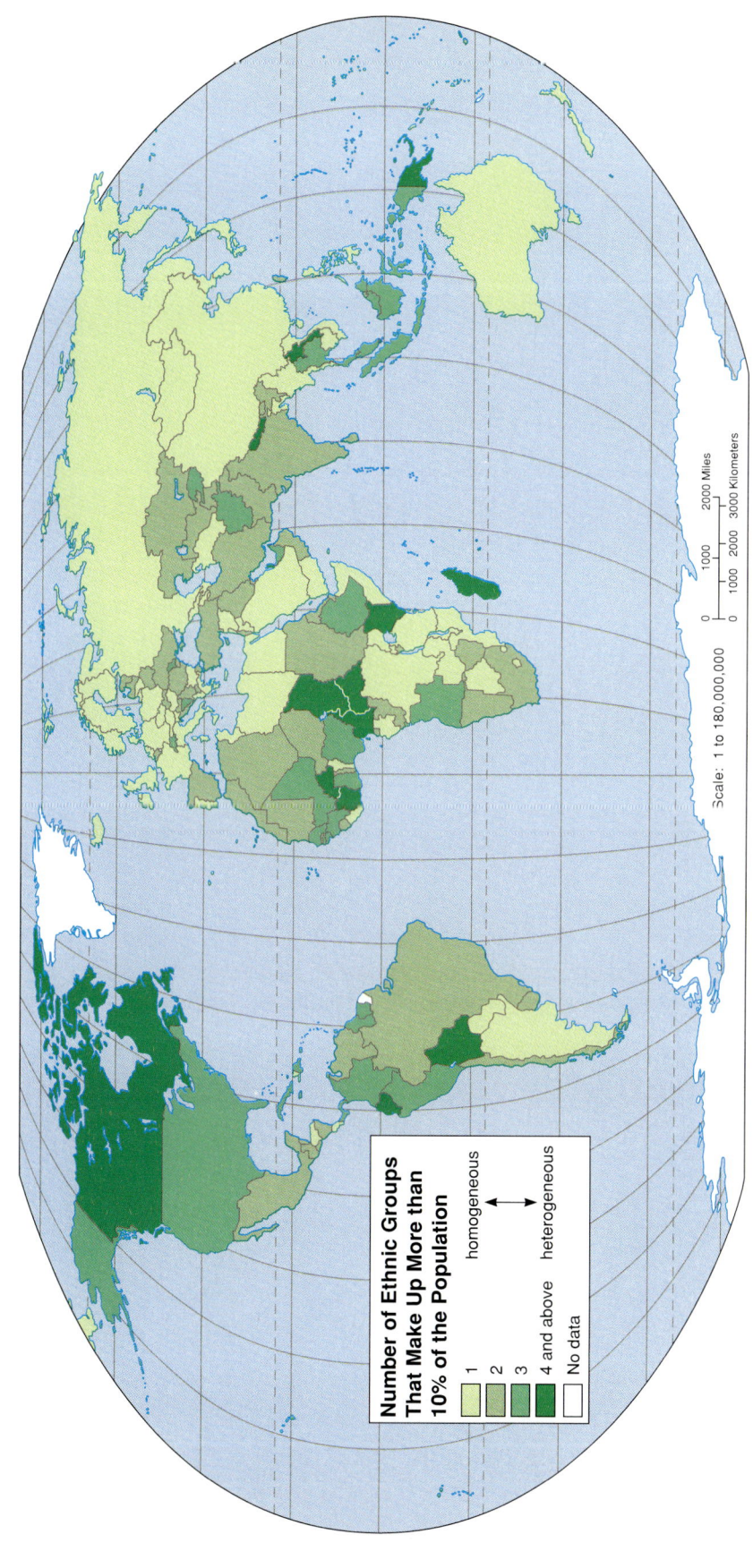

Number of Ethnic Groups That Make Up More than 10% of the Population
- 1 (homogeneous)
- 2
- 3
- 4 and above (heterogeneous)
- No data

Scale: 1 to 180,000,000

The presence of minority ethnonational groups within a country's population can add a vibrant and dynamic mix to the whole. Plural societies with a high degree of cultural and ethnic diversity should, according to some social theorists, be among the world's most healthy. Unfortunately, the reality of the situation is quite different from theory or expectation. The presence of significant minority populations contributed to the disintegration of the Soviet Union; the continuing existence of minority populations within the new states that formed when the USSR broke apart threatens the viability and stability of those young political units. In Africa, national boundaries drawn by colonial powers without regard for the geographical distribution of ethnic groups have led to continuing tribal conflicts, which hamper both economic and political development. Even in the most highly developed regions of the world, the presence of minority ethnic populations poses significant problems: witness the separatist movement in Canada, driven by the desire of some French-Canadians to be independent of the English majority, and the continuing ethnic conflict between Flemish-speaking and Walloon-speaking Belgians. This map, which arrays states on a scale of homogeneity to heterogeneity, indicates areas of existing and potential social and political strife.

— 33 —

Map 22 Quality of Life: The Human Development Index

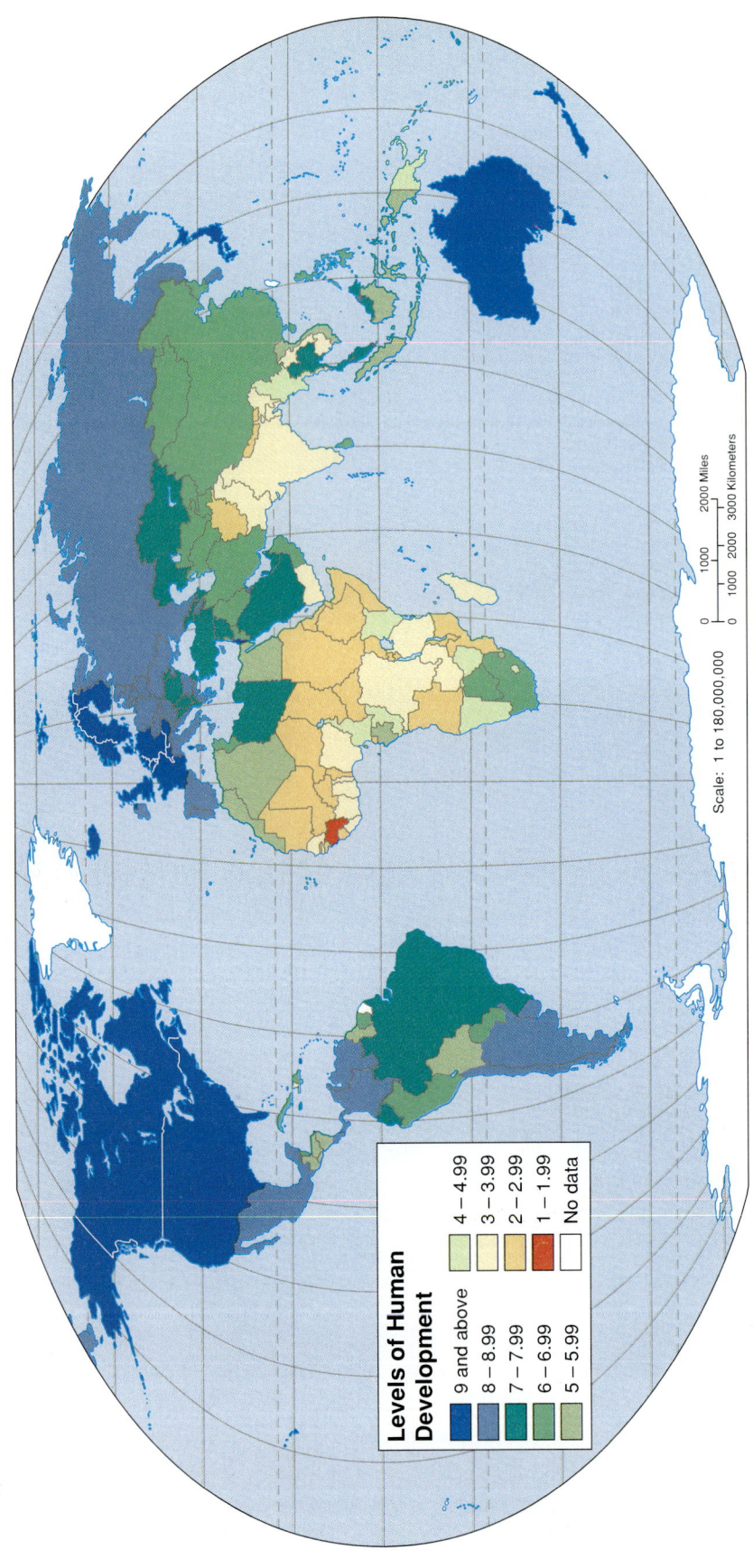

Levels of Human Development

- 9 and above
- 8 – 8.99
- 7 – 7.99
- 6 – 6.99
- 5 – 5.99
- 4 – 4.99
- 3 – 3.99
- 2 – 2.99
- 1 – 1.99
- No data

Scale: 1 to 180,000,000

The development index upon which this map is based takes into account a wide variety of demographic, health, and education data, including population growth, per capita gross domestic product, longevity, literacy, and years of schooling. The map reveals significant improvement in the quality of life in Middle and South America, although whether the gains made in those regions can be maintained in the face of the dramatic population increases expected over the next 30 years is questionable. More clearly than anything else, the map illustrates the near-desperate situation in Africa and South Asia. In those regions, the unparalleled growth in population threatens to overwhelm all efforts to improve the quality of life. In Africa, for example, the population is increasing by 20 million persons per year. With nearly 45 percent of the continent's population aged 15 years or younger, this growth rate will accelerate as these people reach childbearing age. Africa, along with South Asia, faces the very difficult challenge of providing basic access to health care, education, and jobs for a rapidly increasing population. The map also illustrates the striking difference in the quality of life between those who inhabit the world's equatorial and tropical regions and those fortunate enough to live in the temperate zones where the quality of life is significantly higher.

— 34 —

Table B
Size and Growth of Population, 1950–2025

	POPULATION (thousands)				AVERAGE ANNUAL POPULATION CHANGE (percent)[1]			AVERAGE ANNUAL INCREMENT TO THE POPULATION (thousands)[1]		
	1950	1990	1995	2025[2]	1980–1985	1990–1995	2000–2005[2]	1980–1985	1990–1995	2000–2005[2]
WORLD[3]	2,519,748	5,284,832	5,716,426	8,294,341	1.7	1.6	1.4	80,396	86,319	87,270
AFRICA[3]	223,967	632,669	728,074	1,495,772	2.9	2.8	2.6	14,627	19,081	22,690
Algeria	8,753	24,935	27,939	45,475	3.1	2.3	2.0	629	601	650
Angola	4,131	9,194	11,072	26,619	2.6	3.7	3.1	197	376	431
Benin	2,046	4,633	5,409	12,252	2.9	3.1	2.8	106	155	191
Botswana	389	1,276	1,487	2,980	3.5	3.1	2.7	34	42	50
Burkina Faso	3,654	8,987	10,319	21,654	2.5	2.8	2.5	184	266	307
Burundi	2,456	5,503	6,393	13,490	2.8	3.0	2.6	124	178	204
Cameroon	4,466	11,526	13,233	29,173	2.8	2.8	2.8	263	341	465
Central African Republic	1,314	2,927	3,315	6,360	2.3	2.5	2.3	56	78	89
Chad	2,658	5,553	6,361	12,907	2.3	2.7	2.5	108	162	192
Congo	808	2,232	2,590	5,677	2.8	3.0	2.6	51	72	83
Egypt	21,834	56,312	62,931	97,301	2.6	2.2	1.7	1,200	1,324	1,256
Equatorial Guinea	226	352	400	798	7.2	2.6	2.4	19	10	12
Eritrea	1,140	3,082	3,531	7,043	2.5	2.7	2.5	63	90	108
Ethiopia	18,434	47,423	55,053	126,686	2.5	3.0	2.9	954	1,526	1,987
Gabon	469	1,146	1,320	2,697	4.0	2.8	2.5	36	35	39
Gambia	294	923	1,118	2,102	3.0	3.8	2.4	21	39	33
Ghana	4,900	15,020	17,453	37,988	3.6	3.0	2.8	421	487	609
Guinea	2,550	5,755	6,700	15,088	2.2	3.0	2.9	105	189	239
Guinea-Bissau	505	964	1,073	1,978	1.9	2.1	2.1	16	22	27
Ivory Coast	2,776	11,974	14,253	36,817	3.9	3.5	3.2	348	456	576
Kenya	6,265	23,613	28,261	63,360	3.6	3.6	3.0	650	930	1,042
Lesotho	734	1,792	2,050	4,172	3.1	2.7	2.6	45	52	64
Liberia	824	2,575	3,039	7,240	3.2	3.3	3.1	65	93	119
Libya	1,029	4,545	5,407	12,885	4.4	3.5	3.2	149	172	222
Madagascar	4,229	12,571	14,763	34,419	3.2	3.2	3.1	314	438	570
Malawi	2,881	9,367	11,129	22,348	3.2	3.5	2.0	213	352	257
Mali	3,520	9,212	10,795	24,575	2.9	3.2	2.9	210	317	394
Mauritania	825	2,003	2,274	4,443	2.6	2.5	2.5	43	54	68
Mauritius	493	1,057	1,117	1,481	1.0	1.1	1.1	10	12	13
Morocco	8,953	24,334	27,028	40,650	2.4	2.1	1.6	487	539	478
Mozambique	6,198	14,187	16,004	35,139	2.3	2.4	2.8	289	363	577
Namibia	511	1,349	1,540	3,049	2.7	2.7	2.5	30	38	47
Niger	2,400	7,731	9,151	22,385	3.4	3.4	3.2	204	284	371
Nigeria	32,935	96,154	111,721	238,397	2.9	3.0	2.7	2,209	3,113	3,765
Rwanda	2,120	6,986	7,952	15,797	3.2	2.6	2.5	179	193	239
Senegal	2,500	7,327	8,312	18,896	2.8	2.5	2.6	167	197	263
Sierra Leone	1,944	3,999	4,509	8,690	2.0	2.4	2.3	69	102	123
Somalia	3,072	8,677	9,250	21,276	3.2	1.3	3.0	232	115	348
South Africa	13,683	37,066	41,465	70,951	2.5	2.2	2.1	775	880	1,007
Sudan	9,190	24,585	28,098	58,388	2.8	2.7	2.6	556	703	898
Swaziland	264	744	855	1,647	3.0	2.8	2.6	18	22	27
Tanzania	7,886	25,600	29,685	62,894	3.2	3.0	2.6	643	817	956

(Continued on next page)

- 35 -

	POPULATION (thousands)				AVERAGE ANNUAL POPULATION CHANGE (percent)[1]			AVERAGE ANNUAL INCREMENT TO THE POPULATION (thousands)[1]		
	1950	1990	1995	2025[2]	1980–1985	1990–1995	2000–2005[2]	1980–1985	1990–1995	2000–2005[2]
Togo	1,329	3,531	4,138	9,377	2.9	3.2	2.9	83	121	152
Tunisia	3,530	8,080	8,896	13,290	2.6	1.9	1.5	175	163	154
Uganda	4,762	17,949	21,297	48,056	2.8	3.4	2.7	398	670	717
Zaire	12,184	37,436	43,901	104,639	3.2	3.2	3.0	938	1,293	1,651
Zambia	2,440	8,150	9,456	19,130	3.6	3.0	2.4	225	261	274
Zimbabwe	2,730	9,903	11,261	19,631	3.3	2.6	2.0	253	272	267
EUROPE[3]	**548,711**	**721,734**	**726,999**	**718,203**	**0.4**	**0.2**	**0.0**	**2,676**	**1,053**	**17**
Albania	1,230	3,289	3,441	4,688	2.1	0.9	1.1	58	30	41
Austria	6,935	7,705	7,968	8,262	0.0[4]	0.7	0.2	2	53	18
Belarus	7,798	10,212	10,141	9,903	0.7	(0.1)	(0.1)	66	(14)	(12)
Belgium	8,639	9,951	10,113	10,407	0.0	0.3	0.1	1	32	14
Bosnia and Herzegovina	2,661	4,308	3,459	4,474	1.0	(4.4)	0.2	42	(170)	11
Bulgaria	7,251	8,991	8,769	7,768	0.2	(0.5)	(0.4)	20	(44)	(37)
Croatia	3,850	4,517	4,495	4,234	0.4	(0.1)	(0.1)	19	(4)	(6)
Czech Republic	8,925	10,306	10,296	10,622	0.0	(0.0)	0.1	4	(2)	11
Denmark	4,271	5,140	5,181	5,081	(0.0)	0.2	0.0	(2)	8	(1)
Estonia	1,101	1,575	1,530	1,422	0.8	(0.6)	(0.0)	11	(9)	(5)
Finland	4,009	4,986	5,107	5,407	0.5	0.5	0.3	24	24	13
France	41,829	56,718	57,981	61,247	0.5	0.4	0.2	258	253	132
Germany	68,376	79,355	81,591	76,442	(0.2)	0.6	(0.1)	(127)	445	(83)
Greece	7,566	10,238	10,451	9,868	0.6	0.4	0.0	58	43	1
Hungary	9,338	10,365	10,115	9,397	(0.2)	(0.5)	(0.3)	(26)	(50)	(29)
Iceland	143	255	269	337	1.1	1.1	0.9	3	3	3
Ireland	2,969	3,503	3,553	3,882	0.9	0.3	0.4	30	10	16
Italy	47,104	57,023	57,187	52,324	0.1	0.1	(0.2)	67	33	(87)
Latvia	1,949	2,671	2,557	2,335	0.6	(0.9)	(0.5)	16	(23)	(11)
Lithuania	2,567	3,711	3,700	3,816	0.9	(0.1)	0.1	31	(2)	2
Macedonia	1,230	2,046	2,163	2,571	1.4	1.1	0.7	26	23	16
Moldova	2,472	4,362	4,432	5,130	1.0	0.3	0.5	41	14	22
Netherlands	10,114	14,952	15,503	16,276	0.5	0.7	0.3	70	110	47
Norway	3,265	4,241	4,337	4,719	0.3	0.5	0.3	13	19	15
Poland	24,824	38,119	38,388	41,542	0.9	0.1	0.3	326	54	103
Portugal	8,405	9,868	9,823	9,685	0.3	(0.1)	(0.0)	28	(9)	(4)
Romania	16,311	23,207	22,835	21,735	0.5	(0.3)	(0.2)	105	(74)	(41)
Russia	103,283	147,913	147,000	138,548	0.7	(0.1)	(0.2)	910	(183)	(268)
Serbia-Montenegro (Yugoslavia)	7,131	10,156	10,849	11,478	0.7	1.3	0.4	65	139	40
Slovakia	3,463	5,256	5,353	6,014	0.7	0.4	0.4	33	19	24
Slovenia	1,473	1,918	1,946	1,825	0.5	0.3	(0.1)	10	6	(2)
Spain	28,009	39,272	39,621	37,571	0.5	0.2	(0.0)	186	70	(8)
Sweden	7,014	8,559	8,780	9,751	0.1	0.5	0.3	8	44	30
Switzerland	4,694	6,834	7,202	7,786	0.7	1.1	0.5	43	74	34
Ukraine	37,024	51,637	51,330	43,715	0.4	(0.1)	(0.2)	191	(51)	(90)
United Kingdom	50,616	57,411	58,258	61,476	0.1	0.3	0.2	58	169	98

(Continued on next page)

- 36 -

	POPULATION (thousands)				AVERAGE ANNUAL POPULATION CHANGE (percent)[1]			AVERAGE ANNUAL INCREMENT TO THE POPULATION (thousands)[1]		
	1950	1990	1995	2025[2]	1980–1985	1990–1995	2000–2005[2]	1980–1985	1990–1995	2000–2005[2]
NORTH AND MIDDLE AMERICA[3]	**219,633**	**423,658**	**454,229**	**615,549**	**1.3**	**1.4**	**1.1**	**5,097**	**6,114**	**5,492**
Belize	69	189	215	386	2.6	2.6	2.3	4	5	6
Canada	13,737	27,791	29,463	38,266	1.1	1.2	0.9	270	334	295
Costa Rica	862	3,035	3,424	5,608	2.9	2.4	1.8	72	78	73
Cuba	5,850	10,598	11,041	12,658	0.8	0.8	0.5	78	89	54
Dominican Republic	2,353	7,110	7,823	11,164	2.3	1.9	1.4	136	143	126
El Salvador	1,940	5,172	5,766	9,735	0.9	2.2	2.0	43	119	135
Guatemala	2,969	9,197	10,621	21,668	2.8	2.9	2.7	209	285	350
Haiti	3,261	6,486	7,180	13,128	1.8	2.0	2.1	102	139	173
Honduras	1,380	4,879	5,654	10,656	3.2	3.0	2.5	123	155	172
Jamaica	1,403	2,366	2,447	3,301	1.8	0.7	1.0	36	16	26
Mexico	27,740	84,511	93,674	136,594	2.4	2.1	1.5	1,694	1,833	1,597
Nicaragua	1,109	3,676	4,433	9,079	2.8	3.7	2.8	85	151	154
Panama	860	2,398	2,631	3,767	2.1	1.9	1.4	43	47	42
Trinidad and Tobago	636	1,236	1,306	1,808	1.4	1.1	1.1	16	14	16
United States	152,271	249,924	263,250	331,152	0.9	1.0	0.8	2,142	2,665	2,227
SOUTH AMERICA[3]	**111,690**	**293,131**	**319,790**	**462,664**	**2.1**	**1.7**	**1.4**	**5,301**	**5,332**	**5,172**
Argentina	17,150	32,547	34,587	46,133	1.5	1.2	1.1	442	408	416
Bolivia	2,714	6,573	7,414	13,131	1.9	2.4	2.2	108	168	189
Brazil	53,444	148,477	181,790	230,250	2.2	1.7	1.4	2,751	2,663	2,517
Chile	6,068	13,154	14,262	19,775	1.6	1.6	1.2	187	222	195
Colombia	11,946	32,300	35,101	48,359	2.1	1.7	1.3	591	560	526
Ecuador	3,387	10,264	11,460	17,792	2.7	2.2	1.7	228	239	230
Guyana	423	796	835	1,141	0.8	0.9	1.1	6	8	10
Paraguay	1,351	4,317	4,960	9,017	3.3	2.8	2.3	111	129	134
Peru	7,632	21,588	23,780	36,692	2.4	1.9	1.7	439	438	466
Suriname	215	400	423	599	1.2	1.1	1.1	4	5	5
Uruguay	2,239	3,094	3,188	3,691	0.6	0.6	0.6	19	18	18
Venezuela	5,094	19,502	21,644	34,775	2.5	2.3	1.6	409	468	460
ASIA[3]	**1,402,725**	**3,186,446**	**3,457,957**	**4,959,967**	**1.9**	**1.6**	**1.4**	**52,331**	**54,302**	**53,473**
Afghanistan	8,958	15,045	20,141	45,262	(2.0)	5.8	2.7	(309)	1,019	759
Armenia	1,362	3,352	3,599	4,724	1.0	1.4	1.0	32	49	39
Azerbaijan	2,890	7,117	7,558	10,106	1.8	1.2	1.0	100	88	78
Bangladesh	41,783	108,118	120,433	196,128	2.2	2.2	2.0	2,067	2,463	2,874
Bhutan	734	1,544	1,638	3,136	2.1	1.2	2.3	28	19	45
Cambodia	4,346	8,841	10,251	19,686	3.0	3.0	2.3	213	282	284
China	554,760	1,155,305	1,221,462	1,526,106	1.4	1.1	0.8	14,260	13,231	10,543
Georgia	3,728	5,418	5,457	6,122	0.8	0.1	0.4	41	8	21
India	357,561	850,638	935,744	1,392,086	2.2	1.9	1.6	15,866	17,021	17,040
Indonesia	79,538	182,812	197,588	275,598	2.1	1.6	1.3	3,275	2,955	2,884
Iran	16,913	58,946	67,283	123,549	4.4	2.7	2.5	1,932	1,667	1,991
Iraq	5,158	18,078	20,449	42,656	3.3	2.5	2.8	462	474	713
Israel	1,258	4,660	5,629	7,808	1.8	3.8	1.3	71	194	84
Japan	83,625	123,537	125,095	121,594	0.7	0.3	0.1	806	312	159

(Continued on next page)

- 37 -

	POPULATION (thousands)				AVERAGE ANNUAL POPULATION CHANGE (percent)[1]			AVERAGE ANNUAL INCREMENT TO THE POPULATION (thousands)[1]		
	1950	1990	1995	2025[2]	1980–1985	1990–1995	2000–2005[2]	1980–1985	1990–1995	2000–2005[2]
Jordan	1,237	4,259	5,439	12,039	5.4	4.9	3.0	182	236	211
Kazakhstan	6,756	16,670	17,111	21,748	1.1	0.5	0.8	175	88	149
Korea, North	9,726	21,774	23,917	33,386	1.7	1.9	1.3	326	429	337
Korea, South	20,357	42,869	44,995	54,418	1.4	1.0	0.8	536	425	394
Kuwait	152	2,143	1,547	2,805	4.5	(6.5)	2.5	69	(119)	48
Kyrgyzstan	1,749	4,362	4,745	7,128	2.0	1.7	1.5	75	77	82
Laos	1,755	4,202	4,832	9,688	2.3	3.0	2.6	78	136	153
Lebanon	1,443	2,555	3,009	4,424	(0.0)	3.3	1.5	(0)	91	49
Malaysia	6,110	17,891	20,140	31,577	2.6	2.4	1.7	383	450	406
Mongolia	761	2,177	2,410	3,827	2.8	2.0	1.9	49	47	52
Myanmar	17,832	41,813	46,527	75,564	2.1	2.1	1.9	745	943	1,011
Nepal	7,974	19,253	21,918	40,693	2.6	2.6	2.4	420	533	626
Oman	456	1,751	2,163	6,094	4.8	4.2	3.7	59	82	107
Pakistan	39,513	121,933	140,497	284,827	3.7	2.8	2.7	3,438	3,713	4,657
Philippines	20,988	60,779	67,581	104,522	2.5	2.1	1.8	1,270	1,360	1,384
Saudi Arabia	3,201	16,048	17,880	42,651	5.5	2.2	3.1	609	366	720
Singapore	1,022	2,705	2,848	3,355	1.2	1.0	0.6	29	29	19
Sri Lanka	7,678	17,225	18,354	25,031	1.7	1.3	1.1	259	226	221
Syria	3,495	12,348	14,661	33,506	3.5	3.4	3.2	329	463	599
Tajikistan	1,558	5,287	6,101	11,792	2.8	2.9	2.5	121	163	184
Thailand	20,010	55,573	58,791	73,584	1.8	1.1	0.9	882	642	554
Turkey	20,809	56,098	61,945	90,937	2.5	2.0	1.5	1,181	1,169	1,070
Turkmenistan	1,212	3,657	4,099	6,650	2.4	2.3	1.9	72	88	91
United Arab Emirates	70	1,671	1,904	2,958	6.1	2.6	1.8	73	47	39
Uzbekistan	6,376	20,420	22,843	37,678	2.6	2.2	2.0	435	485	527
Vietnam	29,954	66,689	74,545	118,151	2.2	2.2	1.9	1,237	1,571	1,614
Yemen	4,316	11,311	14,501	33,676	3.1	5.0	3.1	276	638	562
OCEANIA[3]	**12,612**	**26,428**	**28,549**	**41,027**	**1.5**	**1.5**	**1.3**	**354**	**424**	**415**
Australia	8,219	16,868	18,088	24,667	1.4	1.4	1.1	214	240	215
Fiji	289	726	784	1,161	2.0	1.5	1.5	13	12	13
New Zealand	1,908	3,360	3,575	4,376	0.8	1.2	0.8	27	43	31
Papua New Guinea	1,613	3,839	4,302	7,532	2.2	2.3	2.1	71	93	109
Solomon Islands	90	320	378	844	3.5	3.3	3.1	9	12	15

[1] Negative numbers are shown in parentheses.
[2] Estimated.
[3] World and regional totals include countries not listed here.
[4] Zero or less than half the unit of measure.

Sources: United Nations Population Division and International Labor Office; *World Resources 1996–97*.

Table C
World Countries: Demography, 1975–1995

	CRUDE BIRTH RATE (births per 1,000 population)		LIFE EXPECTANCY AT BIRTH (years)		TOTAL FERTILITY RATE		PERCENTAGE OF POPULATION IN SPECIFIC AGE GROUPS					
							1975			1995		
	1970–1975	1990–1995	1970–1975	1990–1995	1970–1975	1990–1995	<15	15–65	>65	<15	15–65	>65
WORLD[1]	30.9	25.0	57.9	64.7	4.5	3.1	36.9	57.5	5.6	31.5	62.0	6.5
AFRICA[1]	46.5	41.9	46.0	52.8	6.6	5.8	44.8	52.2	3.1	44.0	52.8	3.2
Algeria	48.0	29.1	54.5	67.1	7.4	3.9	47.6	48.2	4.2	38.7	57.7	3.6
Angola	49.0	51.3	38.0	46.5	6.6	7.2	44.2	52.9	3.0	47.1	50.0	2.9
Benin	49.4	48.7	40.0	47.6	7.1	7.1	44.8	51.6	3.6	47.4	49.7	2.8
Botswana	48.6	37.1	53.2	64.3	6.6	4.9	50.1	47.8	2.1	43.2	54.3	2.4
Burkina Faso	47.8	46.8	41.2	47.4	6.4	6.5	44.0	53.3	2.8	44.9	52.0	3.1
Burundi	44.0	46.0	44.0	50.2	6.8	6.8	45.5	51.1	3.5	46.3	50.8	3.0
Cameroon	45.3	40.7	45.8	56.0	6.3	5.7	43.4	53.1	3.6	44.0	52.4	3.6
Central African Republic	43.1	41.5	43.0	49.4	5.7	5.7	40.6	55.5	3.9	42.7	53.4	4.0
Congo	46.1	44.7	46.7	51.3	6.3	6.3	44.4	52.2	3.5	45.6	51.0	3.4
Egypt	38.4	29.3	52.1	63.6	5.5	3.9	40.0	55.8	4.2	38.0	57.8	4.2
Equatorial Guinea	42.4	43.5	40.5	48.0	5.7	5.9	40.0	55.6	4.4	43.3	52.8	4.0
Eritrea	46.1	43.0	44.3	50.4	6.2	5.8	44.6	52.9	2.5	44.0	53.1	2.9
Ethiopia	49.3	48.5	41.0	47.5	6.8	7.0	45.8	51.6	2.6	46.4	50.8	2.9
Gabon	30.9	37.3	45.0	53.5	4.3	5.3	32.3	61.9	5.8	39.2	55.1	5.8
Gambia	49.2	43.7	37.0	45.0	6.5	5.6	42.0	54.9	2.9	41.3	55.8	2.9
Ghana	45.8	41.7	50.0	56.0	6.6	6.0	45.4	51.9	2.7	45.3	51.8	2.9
Guinea	51.6	50.6	37.3	44.5	7.0	7.0	45.3	52.0	2.6	47.1	50.3	2.6
Guinea-Bissau	41.4	42.7	36.5	43.5	5.4	5.8	38.1	58.2	3.8	41.7	54.2	4.1
Ivory Coast	51.1	49.9	45.4	51.0	7.4	7.4	45.8	51.8	2.4	49.1	48.2	2.6
Kenya	52.9	44.5	51.0	55.7	8.1	6.3	49.1	47.2	3.7	47.7	49.4	2.9
Lesotho	42.4	36.9	50.4	60.5	5.7	5.2	41.7	54.8	3.6	42.1	53.9	4.0
Liberia	48.1	47.3	47.5	55.4	6.8	6.8	44.0	52.3	3.7	46.0	50.4	3.7
Libya	49.0	41.9	52.9	63.1	7.6	6.4	46.0	51.7	2.2	45.4	52.0	2.6
Madagascar	47.2	43.9	46.5	56.5	6.6	6.1	44.7	52.4	2.9	46.1	51.1	2.8
Malawi	56.6	50.5	41.0	45.0	7.4	7.2	47.2	50.6	2.2	46.7	50.5	2.7
Mali	51.0	50.8	38.5	46.0	7.1	7.1	46.0	51.5	2.5	47.4	50.0	2.5
Mauritania	45.0	30.8	43.5	51.5	6.5	5.4	43.3	53.8	3.0	43.1	53.7	3.2
Mauritius	26.1	20.8	62.9	70.2	3.3	2.4	39.7	57.6	2.8	27.7	66.4	5.8
Morocco	45.6	29.1	52.9	63.3	6.9	3.8	47.2	49.2	3.7	36.1	59.8	4.1
Mozambique	45.7	45.2	42.5	46.4	6.5	6.5	43.8	53.0	3.1	44.7	52.0	3.2
Namibia	42.5	37.0	48.8	58.8	6.0	5.3	42.9	53.7	3.4	41.9	54.4	3.7
Niger	59.8	52.5	39.0	46.5	8.1	7.4	46.4	51.2	2.4	48.4	49.2	2.4
Nigeria	46.3	45.4	43.5	50.4	6.5	6.5	44.9	52.6	2.5	45.6	51.7	2.8
Rwanda	52.9	44.1	44.6	47.3	8.3	6.6	48.2	49.4	2.4	46.0	51.5	2.5
Senegal	49.2	43.0	40.3	49.3	7.0	6.1	44.8	52.4	2.8	44.6	52.5	2.9
Sierra Leone	48.9	49.1	35.0	39.0	6.5	6.5	42.5	54.4	3.1	44.2	52.8	3.0
Somalia	50.1	50.2	41.0	47.0	7.0	7.0	45.4	51.6	3.0	47.5	49.8	2.7
South Africa	39.6	31.2	53.9	62.9	5.5	4.1	40.9	55.2	3.8	37.3	58.3	4.4
Sudan	47.0	39.8	43.7	53.0	6.7	5.7	44.4	52.8	2.7	43.8	53.3	2.9
Swaziland	47.5	38.5	47.3	57.5	6.5	4.9	45.6	51.5	2.9	43.0	54.4	2.7

(Continued on next page)

- 39 -

	CRUDE BIRTH RATE (births per 1,000 population)		LIFE EXPECTANCY AT BIRTH (years)		TOTAL FERTILITY RATE		PERCENTAGE OF POPULATION IN SPECIFIC AGE GROUPS					
							1975			1995		
	1970–1975	1990–1995	1970–1975	1990–1995	1970–1975	1990–1995	<15	15–65	>65	<15	15–65	>65
Tanzania	49.6	43.1	46.5	52.1	6.8	5.9	47.9	49.8	2.3	45.9	51.6	2.6
Togo	45.6	44.5	45.5	55.0	6.6	6.6	44.2	52.7	3.1	45.7	51.1	3.2
Tunisia	37.1	25.6	55.5	67.8	6.2	3.2	43.8	52.7	3.5	34.9	60.7	4.4
Uganda	50.3	51.8	46.5	44.9	6.9	7.3	47.4	50.1	2.5	48.8	48.8	2.4
Zaire	47.7	47.5	46.1	52.0	6.3	6.7	45.3	52.0	2.8	48.0	49.1	2.9
Zambia	49.1	44.6	47.3	48.9	6.9	6.0	46.5	50.9	2.6	47.4	50.2	2.4
Zimbabwe	48.6	39.1	51.5	53.7	7.2	5.0	49.0	48.4	2.6	44.1	53.1	2.8
EUROPE[1]	**15.6**	**11.6**	**70.8**	**72.9**	**2.1**	**1.6**	**23.7**	**64.8**	**11.4**	**19.2**	**67.0**	**13.8**
Albania	31.9	23.8	67.7	72.0	4.7	2.9	39.9	55.6	4.5	31.4	63.1	5.5
Austria	13.7	11.9	70.6	76.2	2.0	1.5	23.2	61.9	14.9	17.8	67.3	14.9
Belarus	15.8	12.0	71.5	69.8	2.2	1.7	25.6	64.5	10.0	21.6	65.8	12.6
Belgium	13.6	12.1	71.4	76.4	1.9	1.6	22.2	63.9	13.9	17.8	66.4	15.8
Bosnia-Herzegovina	21.3	13.4	67.4	72.4	2.6	1.6	30.9	63.7	5.5	22.2	70.0	7.8
Bulgaria	16.2	10.3	71.2	71.2	2.2	1.5	22.0	67.1	10.9	18.3	67.1	14.5
Croatia	15.0	11.3	69.6	71.4	2.0	1.7	21.5	67.5	11.0	19.1	68.2	12.8
Czech Republic	17.3	12.9	70.0	71.3	2.2	1.8	22.3	64.8	12.9	19.4	68.1	12.5
Denmark	14.6	12.5	73.6	75.3	2.0	1.7	22.6	64.0	13.4	17.2	67.6	15.2
Estonia	15.4	11.0	70.5	69.3	2.2	1.6	21.8	66.0	12.2	20.6	66.6	12.8
Finland	13.2	13.1	70.7	75.7	1.6	1.9	22.0	67.4	10.6	19.1	66.8	14.1
France	16.3	12.9	72.4	76.9	2.3	1.7	23.9	62.6	13.5	19.6	65.5	14.9
Germany	11.4	9.9	71.0	76.0	1.6	1.3	21.5	63.6	14.8	16.1	68.7	15.2
Greece	15.9	9.9	72.3	77.6	2.3	1.4	23.9	63.9	12.2	16.7	67.4	15.9
Hungary	15.7	11.7	69.3	69.0	2.1	1.7	20.3	67.0	12.6	18.1	67.9	14.0
Iceland	21.0	17.6	74.3	78.2	2.8	2.2	30.3	61.0	9.2	24.5	64.3	11.2
Ireland	22.1	14.7	71.3	75.3	3.8	2.1	31.2	57.8	11.0	24.4	64.3	11.2
Italy	16.1	9.8	72.1	77.5	2.3	1.3	24.2	63.7	12.0	15.1	68.9	16.0
Latvia	14.4	11.5	70.1	69.1	2.0	1.6	21.1	66.2	12.7	20.6	66.1	13.3
Lithuania	16.6	13.5	71.3	70.4	2.3	1.8	25.5	63.4	11.1	21.9	66.0	12.2
Macedonia	23.8	15.5	67.5	71.8	3.0	2.0	30.7	63.1	6.1	24.4	67.5	8.2
Moldova	18.5	16.0	64.8	67.6	2.6	2.1	28.9	64.3	6.8	26.4	64.4	9.3
Netherlands	15.4	13.0	74.0	77.4	2.0	1.6	25.3	63.9	10.8	18.4	68.4	13.2
Norway	16.8	14.2	74.4	76.3	2.3	1.9	23.8	62.5	13.7	19.5	64.7	15.9
Poland	17.8	13.2	70.4	71.1	2.3	1.9	24.0	66.4	9.5	22.9	66.1	11.0
Portugal	19.5	12.0	68.0	74.6	2.8	1.6	27.9	62.2	9.9	18.8	67.0	14.1
Romania	19.3	11.3	69.0	69.9	2.6	1.5	25.2	65.2	9.6	20.4	67.7	11.8
Russia	15.2	10.9	68.2	67.6	2.0	1.5	23.3	67.8	8.9	21.1	66.9	12.1
Serbia-Montenegro (Yugoslavia)	18.5	14.2	68.8	72.0	2.4	2.0	24.6	66.2	9.2	22.0	66.6	11.4
Slovakia	19.7	14.3	70.0	70.9	2.5	1.9	26.2	64.3	9.6	22.9	66.3	10.8
Slovenia	16.7	10.5	69.8	72.6	2.2	1.5	23.7	65.3	11.0	18.3	69.2	12.4
Spain	19.5	9.7	72.9	77.6	2.9	1.2	27.6	62.4	10.0	16.5	68.6	14.9
Sweden	13.6	14.1	74.7	78.2	1.9	2.1	20.7	64.2	15.1	19.0	63.7	17.3
Switzerland	14.2	12.6	73.8	78.0	1.8	1.6	22.4	65.0	12.6	17.7	68.1	14.2
Ukraine	14.9	11.4	70.1	69.4	2.0	1.6	23.0	66.5	10.5	20.1	65.9	14.0
United Kingdom	14.5	13.5	72.0	76.2	2.0	1.8	23.3	62.7	140	19.6	65.0	15.5

(Continued on next page)

- 40 -

	CRUDE BIRTH RATE (births per 1,000 population)		LIFE EXPECTANCY AT BIRTH (years)		TOTAL FERTILITY RATE		PERCENTAGE OF POPULATION IN SPECIFIC AGE GROUPS					
							1975			1995		
	1970–1975	1990–1995	1970–1975	1990–1995	1970–1975	1990–1995	<15	15–65	>65	<15	15–65	>65
NORTH AND MIDDLE AMERICA[1]	22.8	20.2	68.6	74.1	3.1	2.5	31.2	60.5	8.4	27.0	63.5	9.8
Belize	40.2	34.7	67.6	73.6	6.3	4.2	47.0	48.5	4.5	42.3	53.5	4.2
Canada	16.0	15.1	73.1	77.4	2.0	1.9	26.1	65.4	8.4	20.8	67.3	11.8
Costa Rica	31.5	26.3	68.1	76.3	4.3	3.1	42.2	54.4	3.4	35.0	60.4	4.7
Dominican Republic	38.8	27.0	59.9	69.6	5.6	3.1	45.3	51.6	3.0	35.1	60.9	4.0
El Salvador	42.8	33.5	58.7	66.4	6.1	4.0	45.9	51.2	2.9	40.7	55.2	4.1
Guatemala	44.6	38.7	54.0	64.8	6.5	5.4	45.7	51.5	2.8	44.3	52.2	3.5
Honduras	46.9	37.1	54.1	67.7	7.1	4.9	48.0	49.4	2.6	43.8	53.1	3.1
Jamaica	32.5	21.7	68.6	73.6	5.0	2.4	45.2	49.0	5.8	30.8	62.6	6.6
Mexico	42.4	27.7	62.9	70.8	6.4	3.2	46.3	49.8	3.9	35.9	59.9	4.2
Nicaragua	47.2	40.5	55.3	66.7	6.8	5.0	48.0	49.5	2.5	46.0	50.9	3.1
Panama	35.6	25.0	66.4	72.8	4.9	2.9	42.9	52.9	4.3	33.4	61.4	5.2
Trinidad and Tobago	27.0	20.9	65.7	71.6	3.5	2.4	38.0	57.0	4.9	32.3	62.0	5.7
United States	15.7	15.9	71.3	76.0	2.0	2.1	25.2	64.3	10.5	22.0	65.3	12.6
SOUTH AMERICA[1]	32.9	24.8	60.7	68.5	4.6	3.0	39.5	56.3	4.3	32.7	61.8	5.5
Argentina	23.4	20.4	67.2	72.1	3.2	2.8	29.2	63.2	7.6	28.7	61.8	9.5
Bolivia	45.2	35.7	46.7	59.4	6.5	4.8	43.0	53.5	3.4	40.6	55.6	3.8
Brazil	33.6	24.6	59.8	66.3	4.7	2.9	40.1	56.2	3.7	32.3	62.5	5.2
Chile	27.5	21.9	63.5	73.8	3.6	2.5	37.0	57.7	5.3	29.5	63.8	6.6
Colombia	32.6	24.0	61.7	69.3	4.7	2.7	43.1	53.4	3.5	32.9	62.6	4.5
Ecuador	40.6	28.3	58.9	68.8	6.0	3.5	43.8	52.1	4.0	36.4	59.2	4.4
Guyana	35.0	25.1	60.0	65.2	4.9	2.6	44.1	52.2	3.7	32.2	63.7	4.0
Paraguay	36.6	33.0	65.6	70.0	5.7	4.3	44.3	52.2	3.5	40.3	55.9	3.8
Peru	40.5	27.3	55.5	66.0	6.0	3.4	43.2	53.2	3.5	35.1	60.8	4.1
Suriname	34.6	25.3	64.0	70.3	5.3	2.7	47.8	48.6	3.8	35.0	60.0	5.0
Uruguay	21.1	17.1	68.8	72.5	3.0	2.3	27.7	62.7	9.6	24.4	63.3	12.3
Venezuela	35.1	27.4	66.0	71.7	4.9	3.3	43.3	53.6	3.1	36.2	59.7	4.1
ASIA[1]	33.9	25.2	56.3	64.8	5.1	3.0	39.9	55.9	4.2	32.0	62.7	5.3
Afghanistan	51.6	50.2	38.0	43.5	7.1	6.9	43.8	53.9	2.4	40.8	56.4	2.8
Armenia	22.3	20.7	72.5	72.6	3.0	2.6	34.3	59.8	5.8	29.6	63.0	7.4
Azerbaijan	27.0	22.5	69.0	70.6	4.3	2.5	40.0	54.4	5.6	31.3	62.3	5.9
Bangladesh	48.5	35.5	44.9	55.3	7.0	4.4	45.3	50.5	3.6	39.5	57.5	3.1
Bhutan	41.6	39.6	40.7	50.7	5.9	5.9	39.9	56.8	3.2	41.1	55.5	3.5
Cambodia	39.9	43.5	40.3	51.6	5.5	5.3	41.6	55.6	2.8	44.9	52.4	2.6
China	28.3	18.5	63.2	68.5	4.8	2.0	39.5	56.1	4.4	26.4	67.5	6.1
Georgia	18.7	15.9	69.2	72.8	2.6	2.1	28.4	63.1	8.5	23.7	64.8	11.4
India	38.2	29.1	50.3	60.4	5.4	3.8	39.8	56.4	3.8	35.2	60.2	4.6
Indonesia	38.2	24.7	49.3	62.7	5.1	2.9	42.0	54.8	3.2	33.0	62.7	4.3
Iran	44.1	35.5	55.9	67.5	6.5	5.0	45.4	51.3	3.3	43.5	52.6	3.9
Iraq	47.4	38.1	57.0	66.0	7.1	5.7	46.6	50.9	2.5	43.6	53.5	3.0
Israel	27.4	21.2	71.6	76.5	3.8	2.9	32.9	59.4	7.8	29.1	61.4	9.5
Japan	19.2	10.1	73.3	79.5	2.1	1.5	24.3	67.8	7.9	16.2	69.6	14.1
Jordan	50.0	38.9	56.6	67.9	7.8	5.6	47.2	50.0	2.8	43.3	54.0	2.7

(Continued on next page)

- 41 -

	CRUDE BIRTH RATE (births per 1,000 population)		LIFE EXPECTANCY AT BIRTH (years)		TOTAL FERTILITY RATE		PERCENTAGE OF POPULATION IN SPECIFIC AGE GROUPS					
							1975			1995		
	1970–1975	1990–1995	1970–1975	1990–1995	1970–1975	1990–1995	<15	15–65	>65	<15	15–65	>65
Kazakhstan	25.9	19.8	64.4	69.6	3.5	2.5	34.6	59.7	5.7	29.8	63.2	7.0
Korea, North	35.8	24.1	61.5	71.1	5.7	2.4	45.1	51.7	3.1	29.1	66.3	4.6
Korea, South	28.8	16.4	62.6	71.1	4.1	1.7	37.7	58.6	3.6	23.6	70.8	5.6
Kuwait	44.4	24.2	67.3	74.9	6.9	3.1	44.4	54.0	1.6	39.8	58.5	1.7
Kyrgyzstan	31.1	29.0	63.1	69.0	4.7	3.7	39.9	54.2	5.9	37.1	57.1	5.8
Laos	44.4	45.2	40.4	51.0	6.2	6.7	42.1	55.3	2.7	44.8	52.2	3.0
Lebanon	32.1	26.9	65.0	68.5	4.9	3.1	41.2	53.9	5.0	34.1	60.4	5.5
Malaysia	34.7	28.8	63.0	70.8	5.2	3.6	42.1	54.2	3.7	38.0	58.1	3.9
Mongolia	41.5	27.6	53.8	63.7	5.8	3.6	43.7	53.4	2.9	38.0	58.5	3.4
Myanmar	39.9	32.5	49.8	57.6	5.8	4.2	40.7	55.4	3.8	37.4	58.5	4.1
Nepal	45.6	39.2	43.3	53.5	6.3	5.4	42.3	54.5	3.2	42.4	54.2	3.4
Oman	49.6	43.6	49.0	69.6	7.2	7.2	44.6	52.5	2.7	47.5	49.9	2.6
Pakistan	47.5	40.9	50.6	61.5	7.0	6.2	45.5	51.6	3.0	44.3	52.7	3.0
Philippines	38.4	30.4	57.8	66.3	5.5	3.9	43.6	53.8	2.7	38.3	58.3	3.4
Saudi Arabia	47.6	35.1	53.9	69.7	7.3	6.4	44.3	52.7	3.0	41.9	55.4	2.7
Singapore	21.2	15.9	69.5	74.8	2.6	1.7	32.8	63.1	4.1	22.7	70.5	6.7
Sri Lanka	28.9	20.7	65.0	71.9	4.0	2.5	39.4	56.6	4.1	30.7	63.5	5.8
Syria	46.6	41.1	57.0	67.1	7.7	5.9	48.5	47.8	3.7	47.3	49.9	2.8
Tajikistan	39.7	36.8	63.4	70.2	6.8	4.9	45.4	49.9	4.7	43.1	52.6	4.3
Thailand	35.1	19.4	59.6	69.0	5.0	2.1	44.9	52.1	3.0	28.3	66.7	5.0
Turkey	34.5	27.3	57.9	66.5	5.0	3.4	40.1	55.4	4.5	33.9	61.1	5.0
Turkmenistan	37.1	31.9	60.6	65.0	6.2	4.0	43.5	52.1	4.5	39.5	56.4	4.2
United Arab Emirates	33.0	23.2	62.5	73.8	6.4	4.2	28.1	69.7	2.0	31.3	67.0	1.7
Uzbekistan	35.7	31.5	64.1	69.2	6.0	3.9	43.3	51.1	5.5	39.9	55.6	4.4
Vietnam	37.6	30.7	50.3	65.2	5.9	3.9	43.7	52.3	4.0	37.5	57.7	4.9
Yemen	53.2	49.4	42.1	50.2	7.6	7.6	50.9	46.5	2.6	46.7	50.9	2.4
OCEANIA[1]	**23.9**	**19.2**	**66.6**	**73.0**	**3.2**	**2.5**	**31.1**	**61.5**	**7.5**	**26.0**	**64.4**	**9.5**
Australia	19.6	14.8	71.7	77.6	2.5	1.9	27.6	63.7	8.7	21.6	66.8	11.6
Fiji	32.5	23.7	65.1	71.5	4.2	3.0	39.9	57.5	2.6	34.6	61.5	3.8
New Zealand	20.8	17.3	71.7	75.5	2.8	2.2	30.0	61.3	8.7	23.4	65.3	11.3
Papua New Guinea	41.0	33.4	47.7	55.8	6.1	5.1	42.0	54.9	3.1	39.5	57.5	2.9
Solomon Islands	47.2	37.5	62.0	70.4	7.2	5.4	47.9	46.9	3.2	44.2	52.9	2.9

[1] World and regional totals include countries not listed here.

Sources: United Nations Population Division; World Resources 1996–97.

Table D
Mortality, Health, and Nutrition, 1970–1995

	CRUDE BIRTH RATE PER 1,000 PERSONS		INFANT MORTALITY RATE PER 1,000 LIVE BIRTHS		UNDER-5 MORTALITY RATE PER 1,000 LIVE BIRTHS			POPULATION PER PHYSICIAN		POPULATION PER NURSING PERSON		TOTAL EXPENDITURE ON HEALTH AS A % OF GDP	CALORIES AVAILABLE AS A PERCENT OF NEED
	1970–1975	1990–1995	1970–1975	1990–1995	1960	1980	1993	1970	1993	1970	1993		
WORLD	11.7	9.3	93	64									
AFRICA	**19.2**	**13.7**	**131**	**93**									
Algeria	15.4	6.4	132	55	243	145	68	8,100	X¹	X	X	7.0	123
Angola	26.0	19.2	173	124	345	261	292	X	X	X	X	X	8
Benin	25.7	17.6	135	86	210	176	144	28,960	X	2,610	X	4.3	104
Botswana	13.6	6.6	86	43	170	94	56	15,540	X	1,920	X	X	97
Burkina Faso	24.6	18.2	173	130	318	218	175	96,690	X	X	X	8.5	94
Burundi	20.2	15.7	137	102	255	193	178	58,570	17,240	6,910	4,800	3.3	84
Cameroon	19.5	13.2	119	63	264	173	113	29,390	12,000	2,610	2,000	2.6	95
Central African Republic	21.8	16.7	132	102	294	202	177	44,020	X	2,480	X	4.2	82
Chad	24.9	16.0	166	122	325	254	206	61,900	29,410	8,020	X	6.3	73
Congo	18.9	14.9	95	84	220	125	109	9,940	X	810	X	X	103
Egypt	16.3	8.1	150	67	258	180	59	2,030	1,340	2,480	500	2.6	132
Equatorial Guinea	24.1	18.0	157	117	316	243	X	X	X	X	X	X	X
Eritrea	19.7	15.2	136	105	X	X	204	X	X	X	X	X	X
Ethiopia	22.9	18.0	154	119	260	260	204	85,690	X	X	X	3.6	73
Gabon	20.2	15.5	132	94	194	194	154	5,520	X	570	X	X	104
Gambia	26.7	18.8	179	132	276	X	X	24,420	X	X	X	X	X
Ghana	15.8	11.7	107	81	157	170	170	12,910	X	690	X	3.5	93
Guinea	26.8	20.3	177	134	337	278	226	50,650	X	3,370	X	3.9	97
Guinea-Bissau	26.7	21.3	183	140	336	290	235	17,500	X	2,860	X	X	97
Ivory Coast	19.4	15.1	129	92	300	180	120	15,540	X	1,930	X	3.3	111
Kenya	17.3	11.7	98	69	202	112	90	8,000	X	2,520	X	4.3	89
Lesotho	19.3	10.0	130	79	204	173	156	30,400	X	3,860	X	X	93
Liberia	19.6	14.2	182	126	288	235	217	X	X	X	X	X	98
Libya	14.8	8.1	117	68	269	150	100	X	X	X	X	X	140
Madagascar	19.0	11.8	172	93	364	216	164	10,310	X	2,500	X	2.6	95
Malawi	23.5	30.0	191	143	365	290	223	76,580	50,360	5,330	1,980	5.0	88
Mali	25.4	19.1	203	159	400	210	217	45,320	21,180	2,670	2,050	5.2	96
Mauritania	21.5	14.4	142	101	321	249	202	17,960	X	3,750	X	X	106
Mauritius	7.0	6.6	55	18	84	42	22	4,170	X	610	X	X	128
Morocco	15.7	8.1	122	68	215	145	59	13,090	X	X	X	2.6	125
Mozambique	21.7	18.5	168	148	331	269	262	18,870	X	4,280	X	5.9	77
Namibia	16.8	10.5	113	60	206	114	79	X	4,320	X	X	X	X
Niger	25.2	18.9	167	124	320	320	320	60,360	35,140	5,690	660	5.0	95
Nigeria	20.2	15.4	111	84	204	196	191	20,530	X	4,370	X	2.7	93
Rwanda	20.5	16.7	142	110	191	222	141	60,130	X	5,630	X	2.4	82
Senegal	23.9	16.0	122	68	303	221	120	15,810	X	1,670	X	1.5	98
Sierra Leone	29.2	25.2	193	166	385	301	284	17,830	X	2,700	X	5.6	83
Somalia	23.6	18.5	155	122	294	246	211	X	X	X	X	3.3	81

(Continued on next page)

	MORTALITY					HEALTH				NUTRITION
	CRUDE BIRTH RATE PER 1,000 PERSONS		INFANT MORTALITY RATE PER 1,000 LIVE BIRTHS		UNDER-5 MORTALITY RATE PER 1,000 LIVE BIRTHS			POPULATION PER PHYSICIAN		POPULATION PER NURSING PERSON		TOTAL EXPENDITURE ON HEALTH AS A % OF GDP	CALORIES AVAILABLE AS A PERCENT OF NEED
	1970–1975	1990–1995	1970–1975	1990–1995	1960	1980	1993	1970	1993	1970	1993		
South Africa	13.8	8.8	76	53	91	91	69	X	X	300	X	X	126
Sudan	19.8	13.1	110	78	292	210	128	X	X	X	X	X	87
Swaziland	18.0	10.7	133	75	233	151	X	X	X	X	X	X	X
Tanzania	18.5	13.6	125	85	249	202	167	22,900	3,400	X	X	4.7	X
Togo	19.4	12.8	129	85	164	175	135	28,860	X	1,590	X	4.1	99
Tunisia	12.3	6.4	120	43	244	102	36	5,930	1,540	940	300	4.9	131
Uganda	18.5	19.2	116	115	218	181	185	9,210	X	X	X	3.4	93
Zaire	18.9	14.5	127	93	286	204	187	X	X	X	X	2.4	96
Zambia	18.0	15.1	100	104	220	160	203	13,640	11,430	1,730	610	3.2	87
Zimbabwe	15.1	12.0	93	67	181	125	83	6,310	X	650	X	6.2	94
NORTH AND MIDDLE AMERICA	**9.2**	**7.8**	**35**	**19**									
Belize	7.3	4.8	52	33	X	X	X	X	X	X	X	X	X
Canada	7.4	7.6	16	7	33	13	8	680	X	140	X	9.9	122
Costa Rica	5.8	3.7	53	14	112	29	16	1,620	X	460	X	X	121
Cuba	6.6	6.8	38	12	50	26	10	X	X	X	X	X	135
Dominican Republic	9.9	5.6	94	42	152	94	48	X	X	1,400	X	3.7	102
El Salvador	10.9	7.1	99	46	210	120	60	4,100	X	890	X	5.9	102
Guatemala	13.4	7.7	95	48	205	136	73	3,660	X	X	X	3.7	103
Haiti	17.8	11.9	135	86	270	195	130	X	X	X	X	7.0	89
Honduras	13.4	6.1	104	43	203	100	56	X	X	X	X	4.5	96
Jamaica	8.2	6.2	42	14	76	39	13	2,630	X	530	X	X	114
Mexico	9.2	5.3	68	36	141	81	32	1,480	X	1,620	X	3.2	X
Nicaragua	12.7	6.8	100	52	209	143	72	2,150	1,490	X	X	8.6	99
Panama	7.5	5.3	43	25	104	31	20	1,630	X	1,540	X	X	98
Trinidad and Tobago	7.7	6.1	42	18	73	40	21	2,250	X	150	X	X	114
United States	9.2	8.8	18	9	30	15	10	630	X	160	X	13.3	138
SOUTH AMERICA	**9.7**	**7.1**	**84**	**48**									
Argentina	9.0	8.2	48	24	68	41	27	530	X	960	X	4.2	131
Bolivia	19.0	10.2	151	75	252	170	114	1,970	X	2,990	X	4.0	84
Brazil	9.7	7.5	91	58	181	93	63	2,030	X	4,140	X	4.2	114
Chile	8.9	5.7	69	16	138	35	17	2,160	2,150	460	330	4.7	102
Colombia	8.6	6.0	73	37	132	59	19	2,260	X	X	X	4.0	106
Ecuador	11.5	6.2	95	50	180	101	57	2,870	960	2,640	600	4.1	105
Guyana	10.3	7.1	79	48	128	86	X	X	X	X	X	X	X
Paraguay	7.2	5.5	55	38	90	61	34	2,300	1,260	2,210	X	2.8	116
Peru	12.8	6.9	110	64	236	130	62	1,920	940	X	X	3.2	87
Suriname	7.6	5.8	49	28	96	52	X	X	X	X	X	X	X
Uruguay	10.1	10.3	46	20	47	42	21	910	X	X	X	4.6	101
Venezuela	6.5	4.7	49	23	70	42	24	1,130	640	450	330	3.6	99
ASIA	**11.4**	**8.4**	**98**	**65**									
Afghanistan	26.0	21.8	194	163	360	280	257	X	X	X	X	X	72

(Continued on next page)

MORTALITY

	CRUDE BIRTH RATE PER 1,000 PERSONS		INFANT MORTALITY RATE PER 1,000 LIVE BIRTHS		UNDER-5 MORTALITY RATE PER 1,000 LIVE BIRTHS			
	1970–1975	1990–1995	1970–1975	1990–1995	1960	1980	1993
Armenia	5.7	6.5	22	21	X	X	33
Azerbaijan	6.9	6.4	35	28	X	X	52
Bangladesh	20.8	11.7	140	108	247	211	122
Bhutan	22.6	15.3	178	124	324	249	197
Cambodia	22.5	14.3	181	116	217	330	181
China	6.3	7.2	61	44	209	65	43
Georgia	9.2	8.9	33	19	X	X	28
India	15.8	10.0	132	82	236	177	122
Indonesia	17.3	8.4	114	58	216	128	111
Iran	14.5	6.7	122	36	233	126	54
Iraq	14.6	6.7	96	58	171	83	71
Israel	7.1	6.9	23	9	39	19	9
Japan	6.6	7.6	12	4	40	11	6
Jordan	14.4	5.5	82	36	149	66	27
Kazakhstan	9.2	7.5	50	30	X	X	49
Korea, North	8.3	5.3	47	24	120	43	32
Korea, South	8.9	6.2	38	11	124	18	9
Kuwait	5.0	2.1	43	18	128	35	13
Kyrgyzstan	10.4	6.9	59	35	X	X	58
Laos	22.7	15.2	145	97	233	190	141
Lebanon	9.3	7.1	48	34	91	52	40
Malaysia	8.8	5.1	42	13	105	42	17
Mongolia	13.1	7.4	98	60	185	112	78
Myanmar	16.1	11.1	122	84	237	146	111
Nepal	21.1	13.3	153	99	279	177	128
Oman	20.0	4.8	145	30	300	95	29
Pakistan	17.7	8.3	140	91	221	151	137
Philippines	10.2	6.4	71	44	102	70	59
Saudi Arabia	16.9	4.7	105	29	292	90	38
Singapore	5.1	5.7	19	6	40	13	6
Sri Lanka	8.1	5.8	56	18	130	52	19
Syria	12.1	5.8	88	39	201	73	39
Tajikistan	9.8	6.1	74	48	X	X	83
Thailand	9.3	6.1	65	37	146	61	33
Turkey	11.6	7.4	138	65	217	141	84
Turkmenistan	10.3	7.6	78	57	X	X	89
United Arab Emirates	9.9.	2.7	57	19	240	64	21
Uzbekistan	9.2	6.2	63	41	X	X	66
Vietnam	14.3	8.0	106	42	219	105	48
Yemen	22.1	15.5	184	119	X	X	X
EUROPE	**10.1**	**11.2**	**25**	**12**			
Albania	6.9	5.8	58	30	151	57	41
Austria	12.8	10.8	24	7	43	17	8
Belarus	8.8	11.6	21	16	X	X	22

HEALTH

	POPULATION PER PHYSICIAN		POPULATION PER NURSING PERSON		TOTAL EXPENDITURE ON HEALTH AS A % OF GDP
	1970	1993	1970	1993	
Armenia	350	260	140	100	X
Azerbaijan	390	260	130	110	X
Bangladesh	8,450	5,220	65,810	11,350	3.2
Bhutan	X	X	X	X	X
Cambodia	X	X	X	X	X
China	1,500	1,060	2,500	1,490	3.5
Georgia	280	180	110	80	X
India	4,950	X	3,760	X	6.0
Indonesia	27,440	X	4,910	X	2.0
Iran	3,270	X	1,780	X	2.6
Iraq	X	X	X	X	X
Israel	410	X	X	X	4.2
Japan	890	X	310	X	6.8
Jordan	2,480	770	870	500	3.8
Kazakhstan	460	250	120	90	X
Korea, North	X	X	X	X	X
Korea, South	2,220	950	1,190	450	6.6
Kuwait	1,050	X	260	X	X
Kyrgyzstan	480	310	140	110	X
Laos	15,160	4,450	1,380	490	2.5
Lebanon	X	X	X	X	X
Malaysia	4,310	2,410	1,270	470	3.0
Mongolia	580	360	250	X	X
Myanmar	8,820	12,900	3,050	1,240	X
Nepal	52,050	16,110	17,970	2,300	4.5
Oman	9,270	X	3,820	X	X
Pakistan	4,670	X	7,020	X	3.4
Philippines	9,270	X	2,680	X	2.0
Saudi Arabia	7,460	710	2,080	460	4.8
Singapore	1,520	X	280	X	1.9
Sri Lanka	5,900	X	1,290	X	3.7
Syria	X	X	X	X	2.1
Tajikistan	630	430	190	140	X
Thailand	8,290	4,420	1,170	910	5.0
Turkey	2,230	980	1,000	1,110	4.0
Turkmenistan	460	280	140	90	X
United Arab Emirates	1,120	1,100	X	580	X
Uzbekistan	490	280	150	90	X
Vietnam	X	2,300	4,310	400	2.1
Yemen	34,790	X	X	X	X
EUROPE					
Albania	1,070	X	230	X	X
Austria	550	230	300	X	8.5
Belarus	390	230	120	90	X

NUTRITION

	CALORIES AVAILABLE AS A PERCENT OF NEED
Armenia	X
Azerbaijan	X
Bangladesh	88
Bhutan	128
Cambodia	96
China	112
Georgia	X
India	101
Indonesia	121
Iran	125
Iraq	128
Israel	125
Japan	125
Jordan	110
Kazakhstan	X
Korea, North	121
Korea, South	120
Kuwait	X
Kyrgyzstan	X
Laos	111
Lebanon	127
Malaysia	120
Mongolia	97
Myanmar	114
Nepal	100
Oman	X
Pakistan	99
Philippines	104
Saudi Arabia	120
Singapore	136
Sri Lanka	101
Syria	126
Tajikistan	X
Thailand	103
Turkey	127
Turkmenistan	X
United Arab Emirates	X
Uzbekistan	X
Vietnam	103
Yemen	X
EUROPE	
Albania	107
Austria	133
Belarus	X

(Continued on next page)

MORTALITY · HEALTH · NUTRITION

	CRUDE BIRTH RATE PER 1,000 PERSONS 1970–1975	CRUDE BIRTH RATE PER 1,000 PERSONS 1990–1995	INFANT MORTALITY RATE PER 1,000 LIVE BIRTHS 1970–1975	INFANT MORTALITY RATE PER 1,000 LIVE BIRTHS 1990–1995	UNDER-5 MORTALITY RATE PER 1,000 LIVE BIRTHS 1960	UNDER-5 MORTALITY RATE PER 1,000 LIVE BIRTHS 1980	UNDER-5 MORTALITY RATE PER 1,000 LIVE BIRTHS 1993	POPULATION PER PHYSICIAN 1970	POPULATION PER PHYSICIAN 1993	POPULATION PER NURSING PERSON 1970	POPULATION PER NURSING PERSON 1993	TOTAL EXPENDITURE ON HEALTH AS A % OF GDP	CALORIES AVAILABLE AS A PERCENT OF NEED
Belgium	12.1	10.9	19	6	35	15	10	650	X	X	X	8.1	149
Bosnia-Herzegovina	6.9	7.0	51	15	X	X	X	X	X	X	X	X	X
Bulgaria	6.7	12.7	25	14	70	25	19	540	X	240	X	5.4	148
Croatia	13.5	11.8	27	9	X	X	X	X	X	X	X	X	X
Czech Republic	13.4	13.1	20	9	X	X	10	X	270	X	X	X	X
Denmark	10.0	12.0	12	7	25	10	7	690	X	X	X	7.0	135
Estonia	11.0	12.9	21	16	X	X	23	300	260	110	130	X	X
Finland	8.5	10.3	12	5	28	9	5	960	X	130	X	8.9	113
France	10.6	9.8	16	7	34	13	9	750	X	270	X	9.1	143
Germany	12.4	11.6	21	6	40	16	7	580	X	X	X	9.1	X
Greece	8.6	9.8	34	10	64	23	10	620	X	990	X	4.8	151
Hungary	11.8	14.6	34	15	57	26	15	510	X	210	X	6.0	137
Iceland	7.0	7.0	12	5	22	9	X	X	X	X	X	8.3	X
Ireland	11.0	9.0	18	7	36	14	7	980	X	160	X	8.0	157
Italy	9.8	9.9	26	8	50	17	9	550	X	X	X	8.3	139
Latvia	11.4	13.2	21	14	X	X	26	X	280	X	120	X	X
Lithuania	8.9	11.3	22	13	X	X	20	360	230	130	90	X	X
Macedonia	7.8	7.3	74	27	X	X	9	X	430	X	X	X	X
Moldova	10.2	10.5	37	25	X	X	36	490	250	X	X	X	X
Netherlands	8.3	8.8	12	7	22	11	8	800	X	300	X	8.7	114
Norway	10.0	10.9	12	8	23	11	8	720	X	160	X	8.4	120
Poland	8.3	10.5	27	15	70	24	15	700	450	260	190	5.1	131
Portugal	10.5	10.5	45	10	112	31	11	1,110	X	860	X	6.2	136
Romania	9.4	11.1	40	23	82	36	29	840	540	430	X	3.9	116
Russia	9.1	12.4	28	21	X	X	31	340	220	110	90	X	X
Serbia-Montenegro (Yugoslavia)	9.3	9.6	47	20	27	X	X	X	X	X	X	X	X
Slovakia	10.6	10.6	24	12	X	X	18	X	290	X	110	X	X
Slovenia	10.2	10.8	22	8	X	X	X	X	X	X	X	X	X
Spain	8.3	9.0	21	7	57	16	9	750	X	X	X	6.5	141
Sweden	10.4	11.3	10	5	20	9	6	730	X	140	X	8.8	111
Switzerland	9.0	9.4	13	6	27	11	8	700	X	X	X	8.0	136
Ukraine	9.2	13.2	22	16	X	X	25	360	220	110	90	X	X
United Kingdom	11.7	11.4	17	7	27	14	8	810	X	240	X	6.6.	130
OCEANIA	**9.6**	**7.8**	**41**	**27**									
Australia	8.5	7.4	17	7	24	13	8	840	X	X	X	8.6	124
Fiji	6.2	4.5	45	23	97	42	X	X	X	X	X	X	X
New Zealand	8.4	8.4	16	9	26	16	9	870	X	150	X	7.7	131
Papua New Guinea	17.1	10.7	100	68	248	95	95	11,640	12,750	1,710	1,160	4.4	114
Solomon Islands	9.2	4.4	61	27	X	X	X	X	X	X	X	X	X

[1]X = Not Available

Sources: United Nations Children's Fund; United Nations Development Program; UNESCO; *World Resources 1996–97; World Almanac and Book of Facts 1996.*

Table E
Education and Literacy, 1970–1993

	LITERACY				EDUCATION									
	ADULT FEMALE LITERACY (%)		ADULT MALE LITERACY (%)		PRIMARY SCHOOL ENROLLMENT AS % OF AGE GROUP[1]				SECONDARY SCHOOL ENROLLMENT AS % OF AGE GROUP[1]				HIGHER EDUCATION ENROLLMENT AS % OF AGE GROUP		PRIMARY PUPIL/ TEACHER RATIO
					TOTAL		FEMALE		TOTAL		FEMALE				
	1970	1990	1970	1990	1970	1992	1970	1992	1970	1992	1970	1992	1980	1992	1970	1992
WORLD	X	69	X	82												
AFRICA																
Algeria	X	40	X	62												
Algeria	12	41	39	68	76	99	58	92	11	60	6	53	6	12	X	27
Angola	7	29	16	56	X	X	X	X	X	X	X	X	X	X	X	X
Benin	4	19	17	42	36	66	22	X	5	12	3	7	2	3	41	X
Botswana	31	55	60	78	65	116	67	118	7	54	6	57	1	5	36	29
Burkina Faso	2	7	13	26	13	31	10	24	1	8	1	5	0	1	44	60
Burundi	7	19	30	45	30	69	20	62	2	6	1	4	1	1	37	63
Cameroon	18	44	47	70	89	101	75	93	7	28	4	23	2	3	48	51
Central African Republic	5	41	21	60	64	X	41	X	4	X	2	X	1	2	64	X
Chad	12	29	36	57	35	65	17	41	2	7	0	3	0	1	65	64
Congo	21	59	49	78	X	X	X	X	X	X	X	X	5	6	62	66
Egypt	18	34	47	60	72	101	57	93	35	80	23	73	18	19	38	26
Equatorial Guinea	29	61	65	86	X	X	X	X	X	X	X	X	X	X	X	X
Eritrea	X	X	X	X	X	X	X	X	X	X	X	X	X	X	X	X
Ethiopia	9	21	24	41	16	22	10	18	4	12	2	11	0	1	48	27
Gabon	14	45	39	68	85	X	81	X	8	X	5	X	3	3	46	44
Gambia	7	20	28	48	24	69	15	56	7	18	4	12	0	2	27	30
Ghana	17	46	45	71	64	74	54	67	14	38	8	29	2	2	30	29
Guinea	6	18	25	45	33	42	21	27	13	10	5	5	5	X	X	49
Guinea-Bissau	17	36	43	63	39	X	23	X	8	X	6	X	0	X	X	X
Ivory Coast	6	24	25	44	58	69	45	58	9	24	4	16	3	2	45	37
Kenya	27	62	59	82	58	95	48	93	9	29	5	25	1	X	34	31
Lesotho	34	57	62	78	87	106	101	113	7	25	7	30	2	1	46	51
Liberia	7	18	27	49	X	X	X	X	X	X	X	X	X	X	X	X
Libya	11	54	57	84	X	X	X	X	X	X	X	X	X	X	X	X
Madagascar	43	73	56	88	90	92	82	91	12	X	9	18	3	3	65	38
Malawi	20	37	58	69	X	66	X	60	X	4	X	3	1	X	43	68
Mali	4	17	11	32	22	25	15	19	5	7	2	5	0	1	40	47
Mauritania	17	24	37	47	14	55	8	48	2	14	0	10	X	3	24	51
Mauritius	55	75	76	85	94	106	93	108	30	54	25	56	1	2	32	21
Morocco	9	26	33	52	52	69	36	57	13	28	7	29	6	10	34	28
Mozambique	6	18	26	52	47	60	X	51	5	8	X	5	0	0	X	53
Namibia	X	X	X	X	X	124	X	127	X	41	X	47	0	3	X	32
Niger	1	5	11	18	14	29	10	21	1	6	1	4	0	X	39	38
Nigeria	11	39	32	61	37	76	27	67	4	20	3	17	2	X	34	39
Rwanda	18	44	45	65	68	71	60	70	2	8	1	7	0	X	60	58
Senegal	7	19	24	39	41	58	32	50	10	16	6	X	3	3	45	59
Sierra Leone	5	14	22	40	34	48	27	39	8	16	5	12	1	1	32	X
Somalia	1	14	5	36	X	X	X	X	X	X	X	X	X	X	X	X

(Continued on next page)

EDUCATION

	LITERACY						PRIMARY SCHOOL ENROLLMENT AS % OF AGE GROUP[1]						SECONDARY SCHOOL ENROLLMENT AS % OF AGE GROUP[1]				HIGHER EDUCATION ENROLLMENT AS % OF AGE GROUP		PRIMARY PUPIL/ TEACHER RATIO	
	ADULT FEMALE LITERACY (%)		ADULT MALE LITERACY (%)			TOTAL		FEMALE		TOTAL		FEMALE								
	1970	1990	1970	1990		1970	1992	1970	1992	1970	1992	1970	1992	1980	1992	1970	1992
South Africa	68	79	72	80		99	X	99	X	18	X	17	X	X	X	X	X
Sudan	9	28	33	53		X	X	X	X	X	X	X	X	X	X	X	X
Swaziland	44	71	53	74		X	X	X	X	X	X	X	X	X	X	X	X
Tanzania	21	49	55	75		34	68	27	67	3	5	2	5	0	0	47	36
Togo	10	30	36	61		71	111	44	87	7	23	3	12	2	X	58	59
Tunisia	17	56	42	73		100	117	79	112	23	43	13	42	5	11	47	26
Uganda	22	44	53	70		38	71	30	63	4	13	2	X	1	1	X	X
Zaire	28	61	61	83		X	X	X	X	X	X	X	X	X	X	X	X
Zambia	32	65	64	82		90	97	80	92	13	31	8	26	2	2	47	X
Zimbabwe	58	77	75	88		74	119	66	118	7	48	6	42	1	6	X	38
NORTH AND MIDDLE AMERICA	X	X	X	X													
Belize	X	X	X	X		X	X	X	X	X	X	X	X	X	X	X	X
Canada	X	X	X	X		101	107	100	106	65	104	65	104	42	99	23	17
Costa Rica	87	89	88	94		110	105	109	104	28	43	29	45	23	28	30	32
Cuba	78	94	85	95		X	X	X	X	X	X	X	X	X	X	X	X
Dominican Republic	66	80	70	80		100	X	100	X	21	25	21	27	4	16	55	X
El Salvador	52	67	61	71		85	78	83	79	22	28	22	X	8	X	37	44
Guatemala	36	45	51	60		57	79	51	73	8	28	8	X	X	X	36	34
Haiti	20	38	28	44		X	X	X	X	X	X	X	X	X	X	X	X
Honduras	51	69	57	70		87	105	87	107	14	19	13	34	8	9	35	38
Jamaica	73	87	66	79		119	106	119	108	46	62	45	66	7	9	X	38
Mexico	70	85	80	90		104	113	101	111	22	55	17	55	14	14	46	30
Nicaragua	56	65	58	63		80	102	81	104	18	44	17	46	14	10	37	37
Panama	78	88	80	89		99	106	97	105	38	60	40	46	22	24	27	23
Trinidad and Tobago	89	X	95	98		106	95	107	95	42	81	44	82	5	7	34	26
United States	99	X	99	X		X	104	X	103	X	X	X	X	56	76	23	X
SOUTH AMERICA	X	X	X	X													
Argentina	92	96	93	99		105	107	106	114	44	X	47	X	22	43	19	X
Bolivia	46	71	71	88		76	85	62	81	24	34	20	31	13	23	27	25
Brazil	65	81	72	82		82	106	82	X	26	39	26	X	12	12	28	23
Chile	87	94	89	94		107	96	107	95	39	72	42	75	13	23	50	25
Colombia	79	89	82	90		108	117	110	117	25	55	24	60	10	15	38	28
Ecuador	70	86	79	90		97	X	95	X	22	X	23	X	37	20	37	X
Guyana	88	96	94	98		X	X	X	X	X	X	X	X	X	X	X	X
Paraguay	76	89	86	93		109	110	103	109	17	30	17	31	9	8	32	23
Peru	60	80	83	93		107	119	99	X	31	30	27	X	19	39	35	X
Suriname	76	89	88	94		X	X	X	X	X	X	X	X	X	X	X	X
Uruguay	93	97	92	96		112	108	109	107	69	84	64	X	18	32	24	21
Venezuela	73	89	80	91		94	99	94	100	33	34	34	40	21	30	35	23
ASIA	X	60	X	79													
Afghanistan	3	11	25	42		X	X	X	X	X	X	X	X	X	X	X	X

(Continued on next page)

EDUCATION

LITERACY

	ADULT FEMALE LITERACY (%)		ADULT MALE LITERACY (%)		PRIMARY SCHOOL ENROLLMENT AS % OF AGE GROUP[1]				SECONDARY SCHOOL ENROLLMENT AS % OF AGE GROUP[1]				HIGHER EDUCATION ENROLLMENT AS % OF AGE GROUP		PRIMARY PUPIL/ TEACHER RATIO	
					TOTAL		FEMALE		TOTAL		FEMALE					
	1970	1990	1970	1990	1970	1992	1970	1992	1970	1992	1970	1992	1980	1992	1970	1992
Armenia	X	97	X	99	X	X	X	X	X	X	X	X	X	X	X	X
Azerbaijan	X	96	X	99	X	97	X	X	X	83	X	X	X	X	X	X
Bangladesh	12	23	36	47	54	77	35	71	X	19	X	12	3	4	46	63
Bhutan	9	23	31	51	X	X	X	X	X	X	X	X	X	X	X	X
Cambodia	23	22	X	48	X	X	X	X	X	X	X	X	X	X	X	X
China	36	68	67	87	89	121	X	116	24	51	X	45	1	2	X	22
Georgia	X	98	X	99	X	X	X	X	X	X	X	X	X	X	X	X
India	18	34	48	62	73	102	56	90	26	44	15	32	4	10	41	63
Indonesia	44	75	69	88	80	115	73	113	16	38	11	X	4	10	X	23
Iran	X	56	X	74	72	109	52	104	27	57	18	49	4	12	32	32
Iraq	15	38	44	66	X	X	X	X	X	X	X	X	X	X	X	X
Israel	X	93	X	97	96	94	95	94	57	85	60	89	29	34	16	20
Japan	99	X	99	X	99	102	99	102	86	X	86	X	31	32	26	22
Jordan	34	73	73	91	X	105	X	105	X	X	X	X	27	19	X	X
Kazakhstan	X	96	X	99	X	X	X	X	X	X	X	X	X	X	X	X,
Korea, North	80	95	94	99	103	105	103	106	42	90	32	91	16	42	57	33
Korea, South	45	72	65	78	89	61	76	60	63	51	57	51	11	14	16	16
Kyrgyzstan	X	96	X	99	X	X	X	X	X	X	X	X	X	X	X	X
Laos	18	39	45	65	53	98	40	84	3	22	2	17	1	X	36	29
Lebanon	73	88	87	94	X	X	X	X	X	X	X	X	X	X	X	X
Malaysia	44	74	69	87	87	93	84	94	34	58	28	59	4	7	X	20
Mongolia	52	73	75	87	113	89	X	100	87	77	X	X	X	14	30	X
Myanmar	60	75	84	88	83	105	78	104	21	X	16	X	5	X	47	X
Nepal	4	11	24	37	26	102	8	81	10	36	3	23	3	7	X	39
Oman	X	X	X	X	3	100	1	96	X	57	X	53	0	6	18	27
Pakistan	9	21	31	43	40	46	22	31	13	21	5	13	4	X	41	41
Philippines	83	93	86	94	108	109	X	X	46	74	X	X	28	28	29	36
Saudi Arabia	21	44	49	69	45	78	29	75	12	46	5	41	7	14	24	14
Singapore	61	83	87	95	105	107	101	X	46	X	45	X	8	X	30	X
Sri Lanka	71	85	88	93	99	107	94	105	47	74	48	77	3	6	X	29
Syria	21	49	61	82	X	X	X	X	X	X	X	X	X	X	X	X
Tajikistan	X	96	X	99	X	78	X	X	X	33	X	32	13	19	X	21
Thailand	70	91	86	96	83	97	79	88	17	33	15	32	13	19	35	17
Turkey	40	69	73	90	110	112	94	107	27	60	15	50	6	15	38	29
Turkmenistan	X	97	X	99	X	94	X	X	X	X	X	X	X	X	X	X
United Arab Emirates	40	76	61	77	93	118	71	117	22	69	9	73	2	10	17	17
Uzbekistan	X	96	X	98	X	X	X	X	X	X	X	X	X	X	X	X
Vietnam	65	87	82	95	X	108	X	103	X	33	X	X	2	2	X	X
Yemen	X	X	X	X	22	76	7	37	3	31	0	X	X	X	X	X
EUROPE																
Albania	X	X	X	X	106	101	102	101	35	79	27	74	5	7	X	19
Austria	X	X	X	X	104	103	103	104	72	104	73	100	23	37	21	11
Belarus	X	96	X	99	X	87	X	X	X	91	X	X	X	X	X	X

(Continued on next page)

– 49 –

LITERACY

EDUCATION

	ADULT FEMALE LITERACY (%)		ADULT MALE LITERACY (%)		PRIMARY SCHOOL ENROLLMENT AS % OF AGE GROUP[1]				SECONDARY SCHOOL ENROLLMENT AS % OF AGE GROUP[1]				HIGHER EDUCATION ENROLLMENT AS % OF AGE GROUP		PRIMARY PUPIL/TEACHER RATIO	
					TOTAL		FEMALE		TOTAL		FEMALE					
	1970	1990	1970	1990	1970	1992	1970	1992	1970	1992	1970	1992	1980	1992	1970	1992
Belgium	99	X	99	X	103	99	104	100	81	102	80	103	26	38	20	10
Bosnia-Herzegovina	X	X	X	X	X	X	X	X	X	X	X	X	X	X	X	X
Bulgaria	X	97	X	99	101	90	100	88	79	71	X	73	16	30	22	14
Croatia	X	95	X	99	X	X	X	X	X	X	X	X	X	X	X	X
Czech Republic	X	X	X	X	X	95	X	96	X	88	X	110	X	X	X	18
Denmark	X	X	X	X	96	95	97	95	78	108	75	X	29	38	9	11
Estonia	X	100	X	100	X	85	X	85	X	X	X	X	X	23	X	25
Finland	X	X	X	X	82	100	79	99	102	121	106	133	32	57	22	X
France	98	X	99	X	117	106	117	105	74	101	77	104	26	46	26	12
Germany	X	X	X	X	X	107	X	107	X	X	X	X	27	36	X	16
Greece	76	93	93	98	107	X	106	X	63	X	55	X	17	25	31	19
Hungary	98	98	98	X	97	89	97	89	63	81	55	13	15	18	12	X
Iceland	X	X	X	X	X	X	X	X	X	X	X	X	X	X	X	X
Ireland	X	X	X	X	106	103	106	103	74	101	77	105	20	38	X	25
Italy	99	X	99	X	110	95	109	97	61	76	55	76	28	34	22	12
Latvia	X	99	X	100	X	X	X	X	X	X	X	X	X	23	X	X
Lithuania	X	98	X	99	X	92	X	91	X	X	X	X	X	X	X	X
Macedonia	X	X	X	X	X	X	X	X	X	X	X	X	X	X	X	X
Moldova	X	94	X	99	X	94	X	X	X	X	X	X	X	X	X	20
Netherlands	X	X	X	X	102	98	102	99	75	97	69	96	30	39	30	X
Norway	X	X	X	X	89	99	94	99	83	103	83	104	26	49	20	6
Poland	97	X	99	X	101	98	99	97	62	83	65	86	18	23	23	17
Portugal	65	X	78	X	98	120	96	118	57	68	51	74	11	23	34	14
Romania	X	95	X	98	112	88	113	87	44	80	38	80	11	X	21	21
Russia	X	97	X	99	X	98	X	X	X	X	X	X	X	X	X	X
Serbia-Montenegro (Yugoslavia)	X	89	X	X	X	X	X	X	X	X	X	X	X	X	X	X
Slovakia	X	X	X	X	X	100	X	X	X	96	X	X	X	28	X	22
Slovenia	X	X	X	X	X	X	X	X	X	X	X	X	X	X	X	18
Spain	86	X	94	X	123	107	125	107	56	X	48	93	24	40	34	21
Sweden	X	X	X	X	94	101	95	101	86	91	85	88	31	34	20	10
Switzerland	X	X	X	X	X	105	X	105	X	91	X	88	18	31	X	X
Ukraine	X	97	X	99	X	X	X	X	X	X	X	X	X	X	X	X
United Kingdom	X	X	X	X	104	104	104	105	73	86	73	88	20	28	X	X
OCEANIA																
Australia	X	93	X	96	115	107	115	107	82	82	80	83	25	40	X	17
Fiji	67	86	79	92	X	X	X	X	X	X	X	X	X	X	X	X
New Zealand	X	X	X	X	110	104	109	103	77	84	76	85	29	50	21	16
Papua New Guinea	33	57	60	78	52	73	39	66	8	12	4	10	2	X	30	31
Solomon Islands	X	X	X	X	X	X	X	X	X	X	X	X	X	X	X	X

[1] Large numbers of nontraditional students outside of age group may increase percentages enrolled to above 100 percent in certain countries.

Sources: United Nations Development Program; *Human Development Report 1995*; *World Resources 1996–97*; United Nations Children's Fund; UNESCO; World Health Organization.

– 50 –

Part III

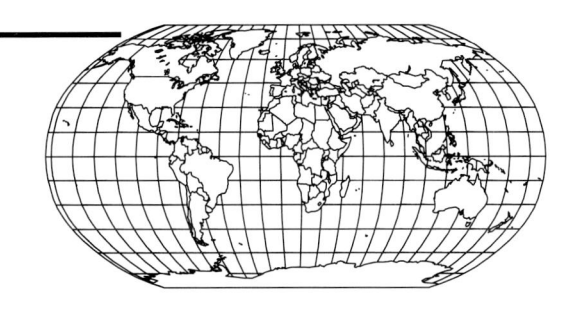

The Economic Factor: The World's Economies

Map 23 Rich and Poor Countries: Gross National Product
Map 24 Gross National Product Per Capita
Map 25 Economic Growth: Change in Gross National Product, 1983-1993
Map 26 Employment by Sector
Map 27 Gross Domestic Product—Share in Agriculture
Map 28 Production of Staples—Cereals, Roots, and Tubers
Map 29 Agricultural Production Per Capita
Map 30 Gross Domestic Product—Share in Exports
Map 31 Exports of Primary Products
Map 32 Dependence on Trade
Map 33 Gross Domestic Product—Share in Investment
Map 34 Relative Wealth of Nations: Purchasing Power Parity
Map 35 Central Government Expenditures Per Capita
Map 36 Energy Production
Map 37 Energy Requirements Per Capita
Map 38 Production of Crucial Materials
Map 39 Composition of Crucial Materials
Table F World Countries: Basic Economic Indicators
Table G World Countries: Agricultural Operations, 1993
Table H World Countries: Energy Production, Consumption, and Requirements, 1993
Table I World Countries: Infrastructure

Map 23 Rich and Poor Countries: Gross National Product

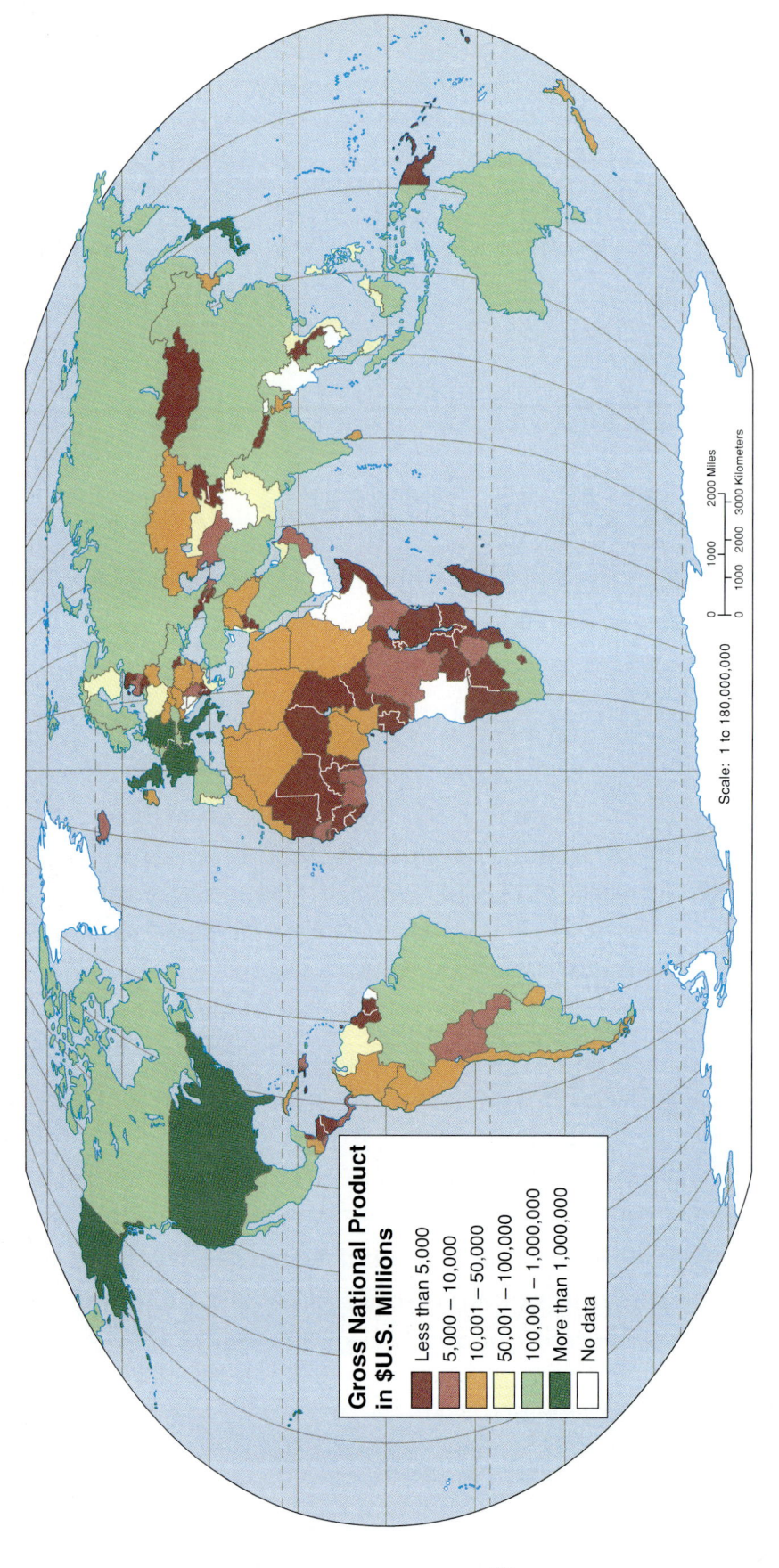

Gross National Product in $U.S. Millions
- Less than 5,000
- 5,000 – 10,000
- 10,001 – 50,000
- 50,001 – 100,000
- 100,001 – 1,000,000
- More than 1,000,000
- No data

Scale: 1 to 180,000,000

Gross National Product (GNP) is the value of all the goods and services produced by a country, including its net income from abroad, during a year. Although GNP is commonly used to measure relative levels of economic development, it is often misleading and incomplete: it does not, for example, take into account environmental deterioration, the accumulation or degradation of human and social capital, or the value of household work. In spite of its deficiencies, however, GNP is still a reasonable way to illustrate the vast differences in wealth that separate the poorest countries from the richest (as long as you keep in mind that GNP provides no measure of the distribution of wealth in a country). One of the more striking features of the map is the evidence that such a small number of countries possess so much of the world's wealth.

— 52 —

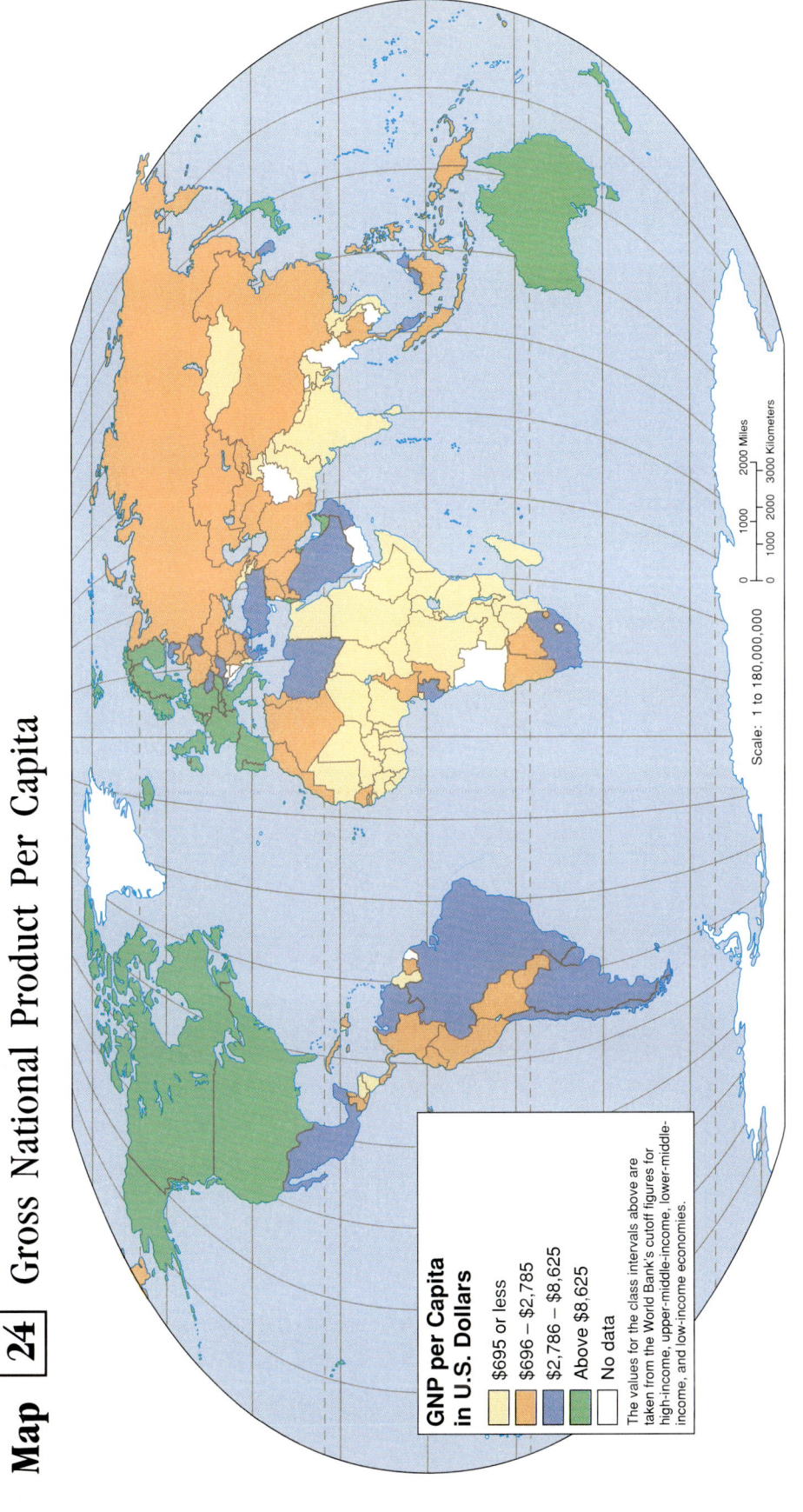

Map 24 Gross National Product Per Capita

GNP per Capita in U.S. Dollars
- $695 or less
- $696 – $2,785
- $2,786 – $8,625
- Above $8,625
- No data

The values for the class intervals above are taken from the World Bank's cutoff figures for high-income, upper-middle-income, lower-middle-income, and low-income economies.

Scale: 1 to 180,000,000

Gross National Product should not be used as the only yardstick of economic development, because it does not measure the distribution of wealth among a population. There are countries (most notably, the oil-rich countries of the Middle East) where per capita GNP is high but where the bulk of the wealth is concentrated in the hands of a few individuals, leaving the remainder in poverty. Even within wealthy countries such as the United States, the distribution of wealth is uneven. The maldistribution tends to be greatest, however, in less developed countries. Still, a low per capita GNP does not automatically condemn a country to low levels of basic human needs and services. There are a few countries (such as Costa Rica and Sri Lanka) that have relatively low per capita GNP figures but rank relatively high in other measures of human well-being, such as average life expectancy, access to medical care, and literacy.

– 53 –

Map 25 Economic Growth: Change in Gross National Product, 1983–1993

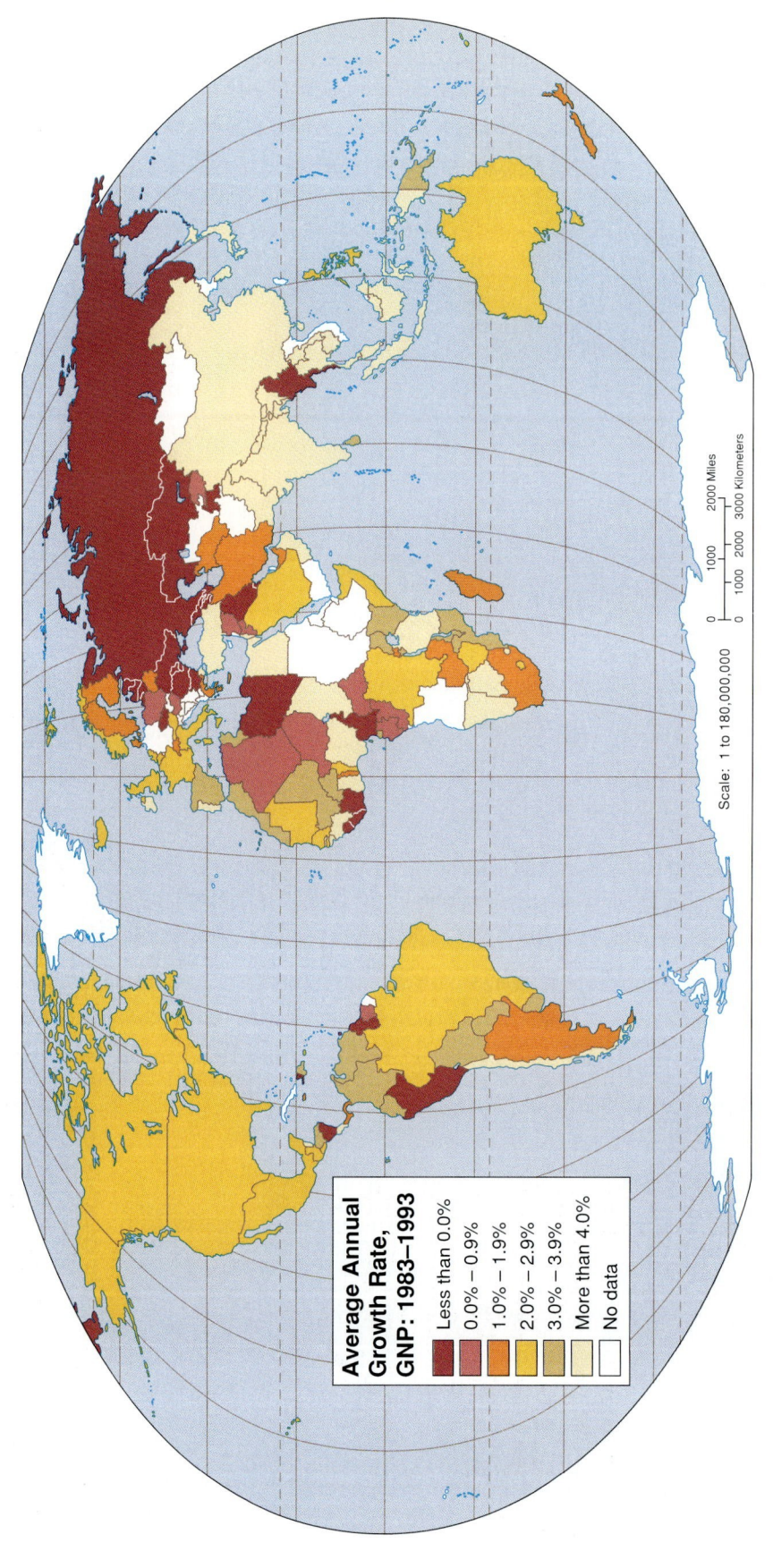

Average Annual Growth Rate, GNP: 1983–1993
- Less than 0.0%
- 0.0% – 0.9%
- 1.0% – 1.9%
- 2.0% – 2.9%
- 3.0% – 3.9%
- More than 4.0%
- No data

Scale: 1 to 180,000,000

As more and more of the world's countries move into the international trade mainstream, the share of the total global production under the control of developing countries increases. During the decade for which these data are available, developing economies in South, Southeast, and East Asia grew at rates far greater than the growth rates of the middle-income and high-income economies of Europe and North and South America. This should not necessarily be viewed as a case of the poor's catching up with the rich; in fact, it shows the huge impact even relatively small production increases will have in countries with small GNPs. Nevertheless, the growth rate in the Gross National Products of some of the world's poorer countries is an encouraging trend.

Map 26 Employment by Sector

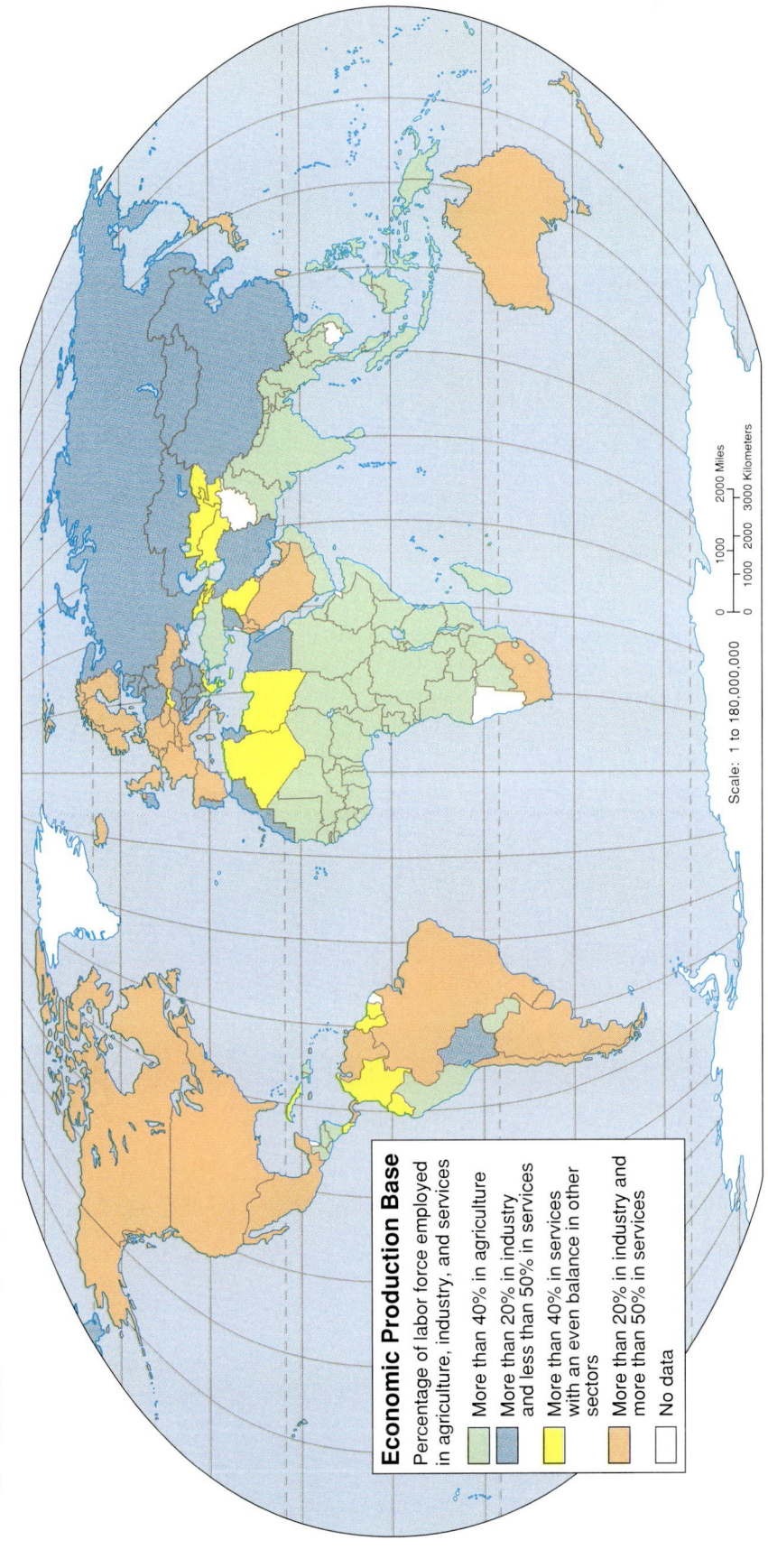

Economic Production Base

Percentage of labor force employed in agriculture, industry, and services

- More than 40% in agriculture
- More than 20% in industry and less than 50% in services
- More than 40% in services with an even balance in other sectors
- More than 20% in industry and more than 50% in services
- No data

Scale: 1 to 180,000,000

The employment structure of a country's population is one of the best indicators of the country's position on the scale of economic development. At one end of the scale are those countries with more than 40 percent of their labor force employed in agriculture. These are almost invariably the least developed, with high population growth rates, poor human services, significant environmental problems, and so on. In the middle of the scale are two types of countries: those with more than 20 percent of their labor force employed in industry and those with a fairly even balance among agricultural, industrial, and service employment but with at least 40 percent of their labor force employed in service activities. These are mostly countries that have undergone the industrial revolution fairly recently and are still developing an industrial base while building up their service activities. They also include countries with a disproportionate share of their economies in service activities, where those service activities are primarily related to resource extraction. On the other end of the scale from the agricultural economies are those countries that have more than 20 percent of their labor force employed in industry and more than 50 percent in service activities. These countries are, for the most part, those with a highly developed, automated industrial base and a highly mechanized agricultural system (the "postindustrial" countries). They also include, particularly where they are found in Latin America and Africa, industrializing countries that are also heavily engaged in resource extraction as a service activity.

– 55 –

Map 27 Gross Domestic Product–Share in Agriculture

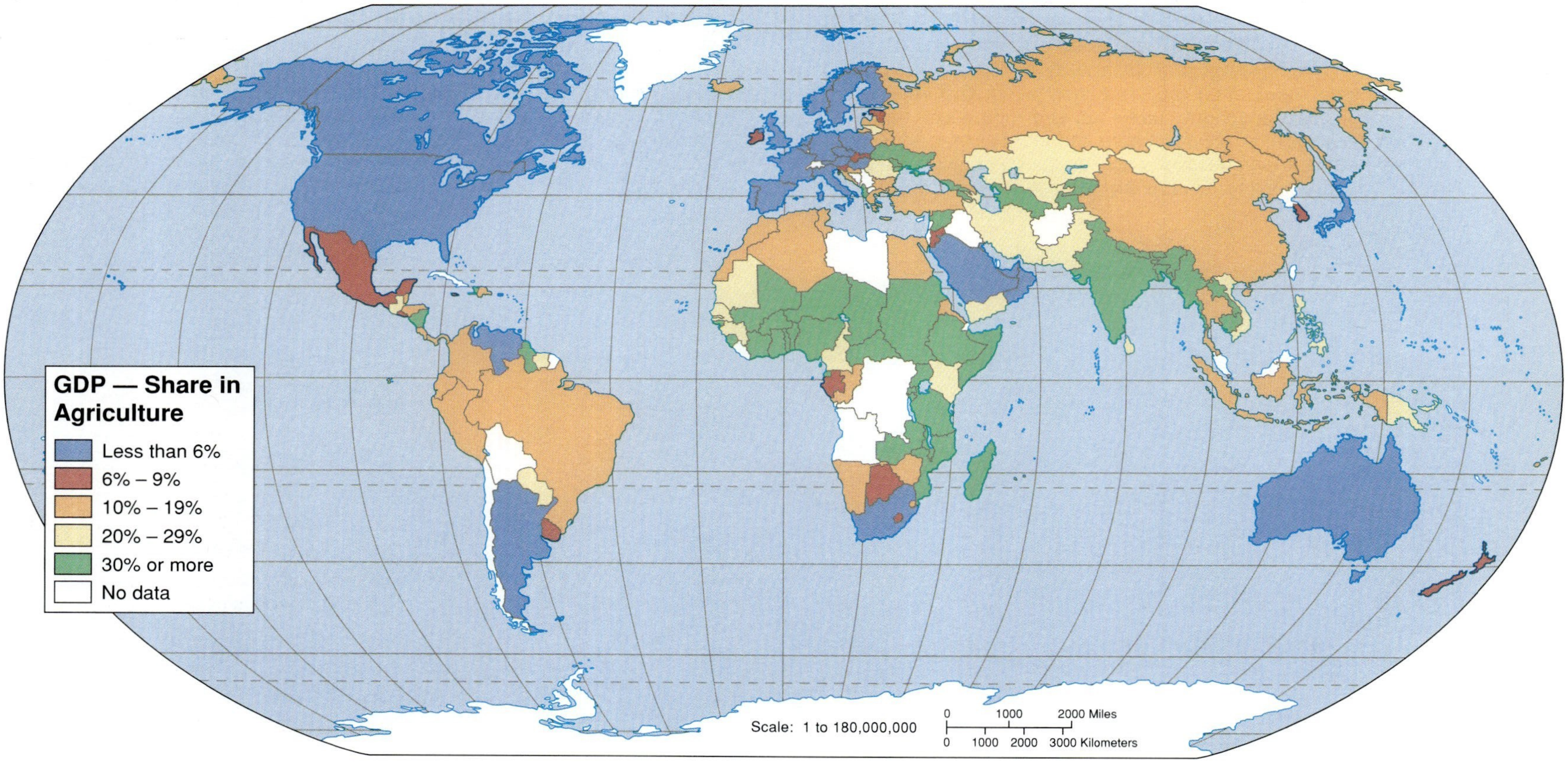

GDP — Share in Agriculture
Less than 6%
6% – 9%
10% – 19%
20% – 29%
30% or more
No data
Scale: 1 to 180,000,000
0 1000 2000 Miles
0 1000 2000 3000 Kilometers

Like employment structures, shares of a country's Gross Domestic Product (GDP) in agriculture, industry, and services are a good indication of the level of economic development. The world's least developed countries have the highest percentages in agriculture; the world's most highly developed countries have the lowest percentages in agriculture and highest percentages in service and industry. This does not mean that those countries with lower shares of GDP in agriculture are unimportant agricultural producers; the United States, Canada, and Australia, for example, have very low percentages of GDP in agriculture but are among the world's most important agricultural producers. What a low GDP share in agriculture most often means is that a country has many other ways of making money. Conversely, a high GDP share in agriculture suggests that a country's economy is limited to agricultural production.

Map 28 Production of Staples—Cereals, Roots, and Tubers

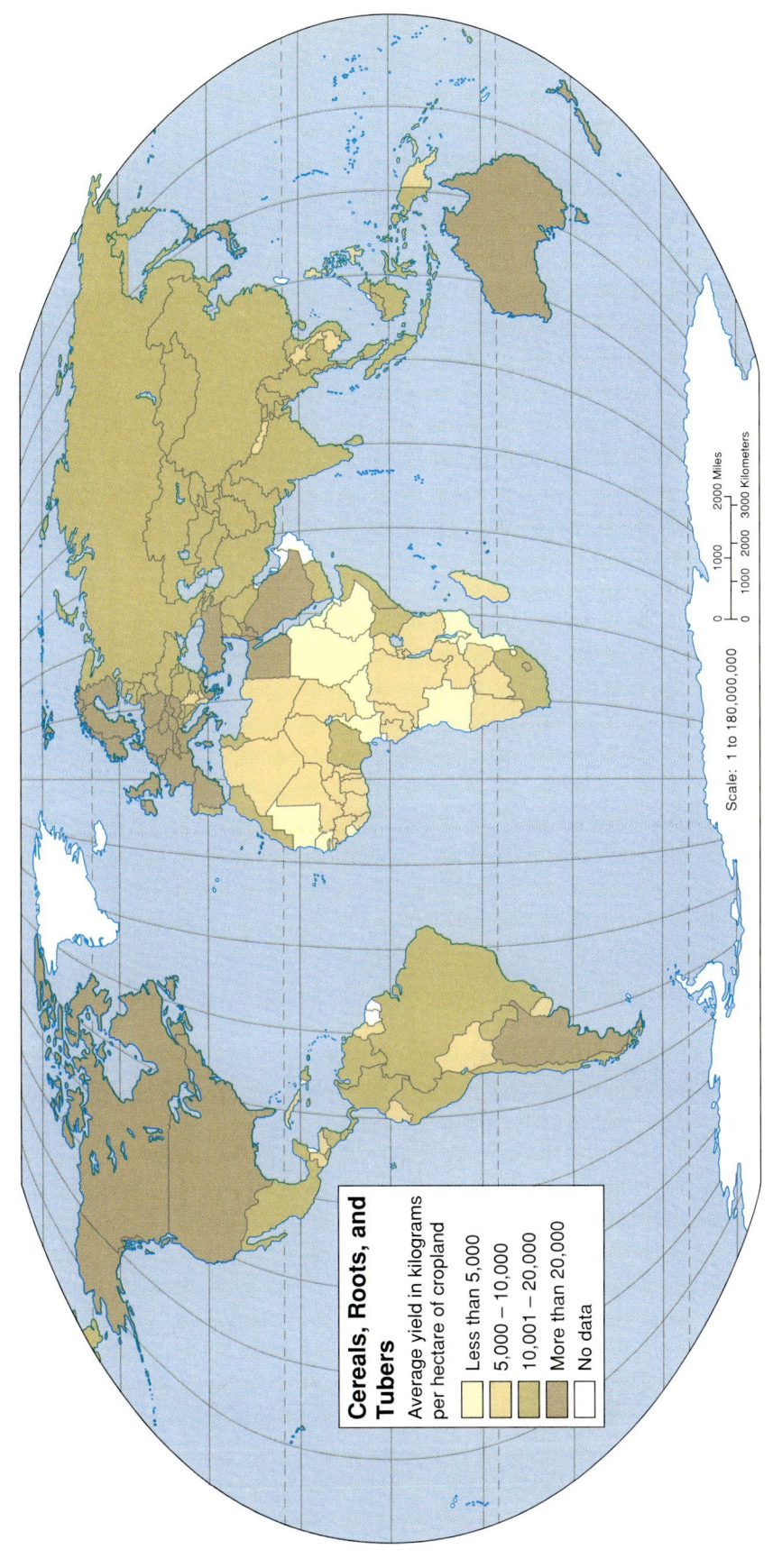

Cereals, Roots, and Tubers
Average yield in kilograms per hectare of cropland
- Less than 5,000
- 5,000 – 10,000
- 10,001 – 20,000
- More than 20,000
- No data

Scale: 1 to 180,000,000

For most of the world's population, the production of crops (as opposed to livestock foods) provides the bulk of dietary intake. Global production of the staple (most important) food crops has increased over the last 10 years—but so has global population. In Africa, for example, despite a 30 percent increase in staple crop production since 1981, per capita food output has dropped more than 5 percent because of population growth that is faster than the growth in agricultural output. The map illustrates considerable regional differences in outputs of food staples per areal unit of cropland. On a global average, 1 hectare (2.47 acres) of cropland in 1990 yielded about 2.6 metric tons (2,600 kilograms, or about 5,700 pounds) of cereals or about 11.8 metric tons of roots and tubers. Yet in Africa, 1 hectare yielded only 1.2 metric tons of cereals or 7.9 metric tons of roots and tubers. In Europe, by contrast, 1 hectare yielded 4.2 and 21.2 metric tons of cereals or roots and tubers, respectively. Such great differences are explainable primarily in terms of agricultural inputs: different farming methodologies, varying levels of fertilizers, agricultural chemicals, irrigation, and machinery. The European farmer applies 2.3 times more fertilizer per hectare than the global average, the African farmer only one-fifth of the global average. These conditions are not likely to change, and the map may be viewed not just as an indicator of present agricultural output but of potential food production as well.

– 57 –

Map 29 Agricultural Production Per Capita

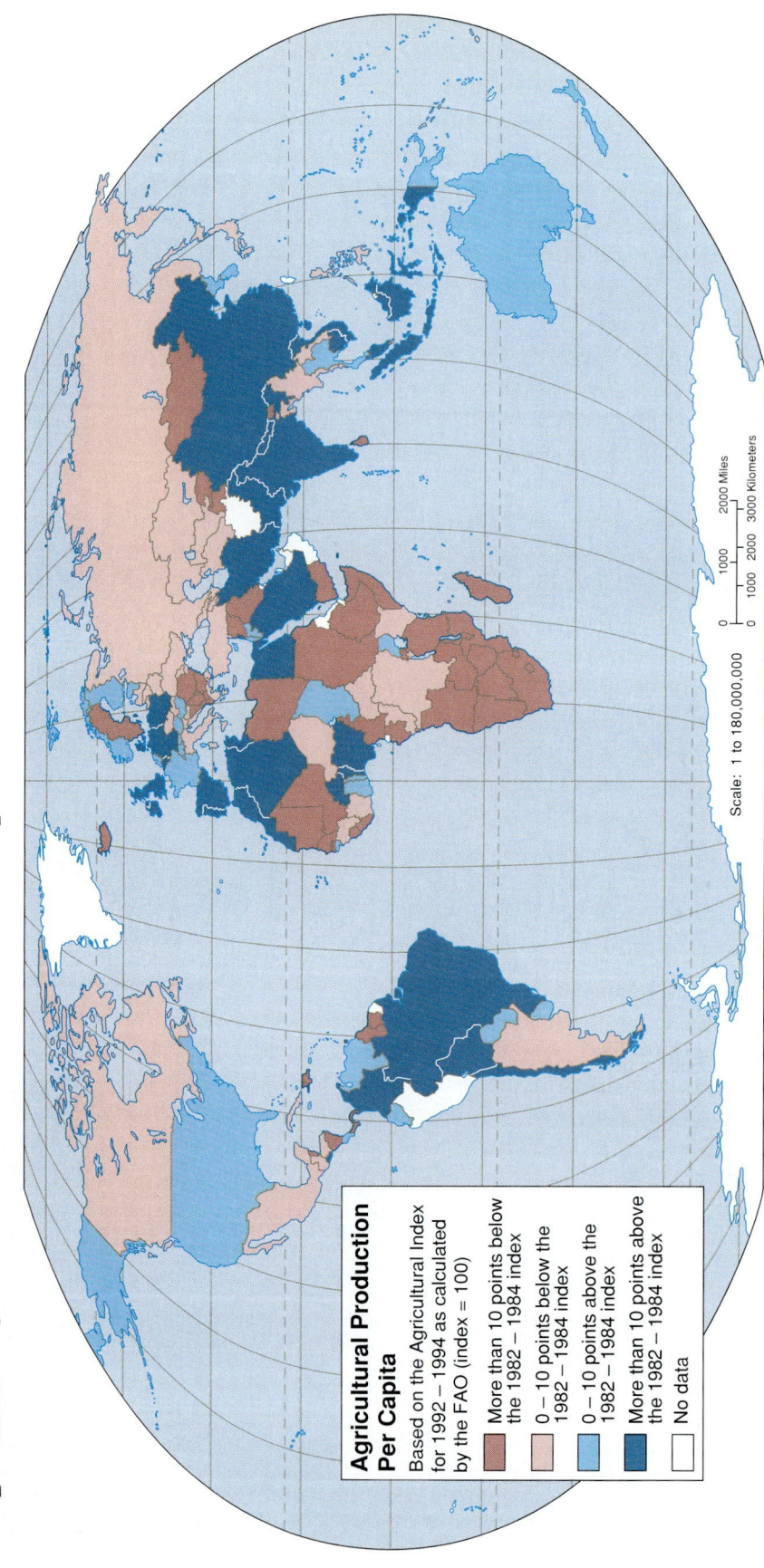

Agricultural Production Per Capita

Based on the Agricultural Index for 1992–1994 as calculated by the FAO (index = 100)

- More than 10 points below the 1982–1984 index
- 0–10 points below the 1982–1984 index
- 0–10 points above the 1982–1984 index
- More than 10 points above the 1982–1984 index
- No data

Scale: 1 to 180,000,000

Agricultural production includes the value of all crop and livestock products originating within a country for the base period of 1992–1994. The index value portrays the disposable output (after deductions for livestock feed and seed for planting) of a country's agriculture in comparison with the base period 1982–1984. Thus, the production values show not only the relative ability of countries to produce food but also whether or not that ability has increased or decreased over the 10-year period from 1982–1984 to 1992–1994. In general, agricultural production in Africa and in Middle America has fallen, while production in South America, Asia, and Europe has risen. In the case of Africa, the drop in production reflects a population growing more rapidly than agricultural productivity. Where rapid increases in food production per capita exist (as in certain countries in South America, Asia, and Europe), most often the reason is the development of new agricultural technologies that have allowed food production to grow faster than population. In much of Asia, for example, the so-called Green Revolution of new, highly productive strains of wheat and rice have made positive index values possible. Also in Asia, the cessation of major warfare has allowed some countries (Cambodia, Laos, Vietnam) to show substantial increases over the 1982–1984 index. In some cases, a drop in production per capita reflects a government policy of limiting production in order to maintain higher prices for agricultural products (as in Japan) or the restructuring of the economic system (as in Russia and many of the other former Soviet republics).

Map 30 Gross Domestic Product–Share in Exports

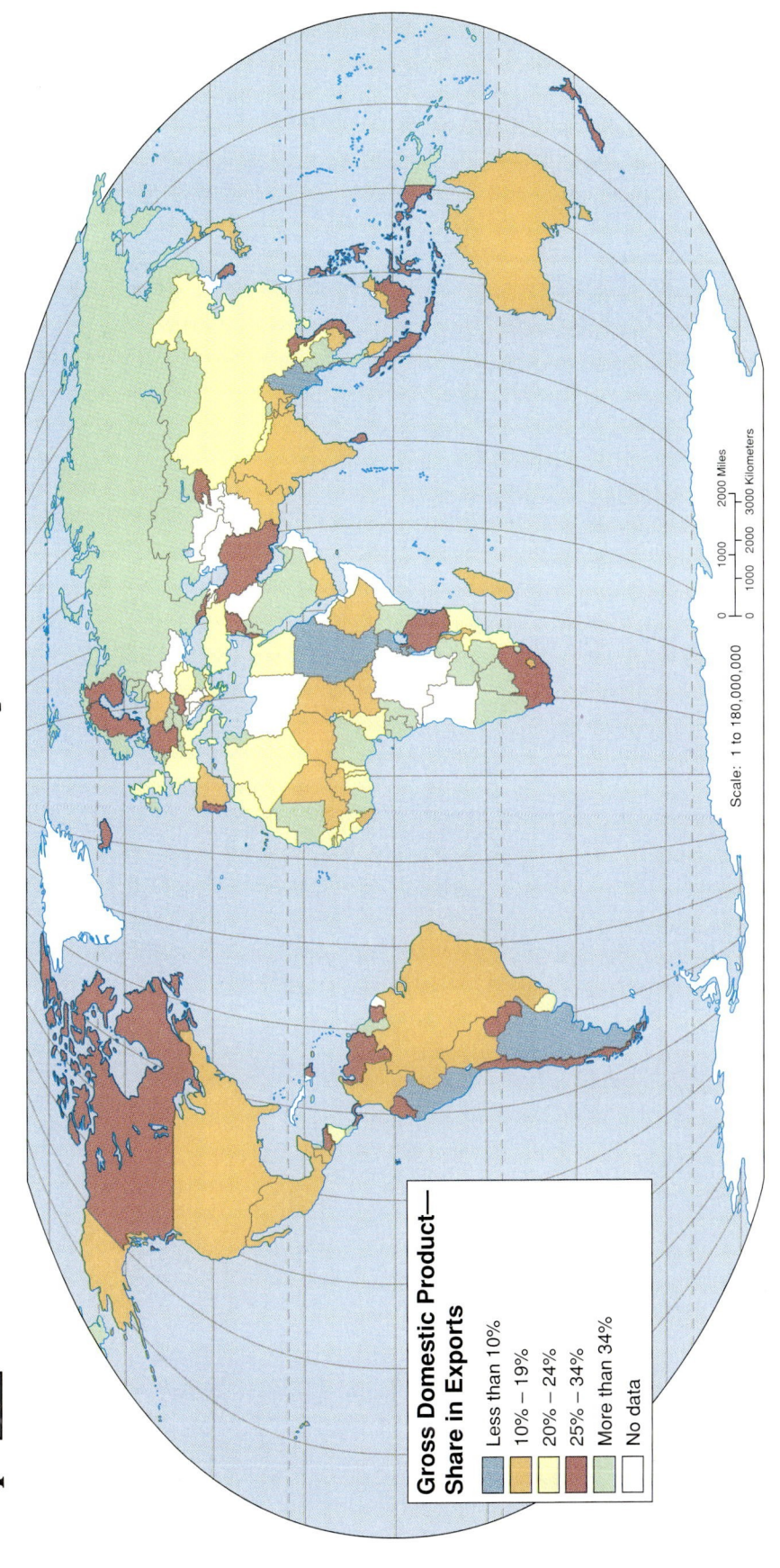

Gross Domestic Product—Share in Exports
- Less than 10%
- 10% – 19%
- 20% – 24%
- 25% – 34%
- More than 34%
- No data

Scale: 1 to 180,000,000

The percentage of GDP consisting of exports is an unusual measure of economic development, often producing similar figures for countries that are otherwise quite different. In general, the countries with the highest percentages of exports in their GDPs are raw material exporters. This includes highly developed economies such as Russia, middle-income countries like Nigeria and other oil exporters, and less developed resource-oriented African countries like Zambia and Zimbabwe. At the other end of the spectrum are countries that have less than 20 percent of their GDP in exports. These range from some of the world's healthiest and most balanced economies on the one hand (the United States, Japan, and Australia, for example) to some of the world's least developed economies and most precarious economic prospects on the other (such as India, Somalia, and Bangladesh). The seeming inconsistencies in the distribution pattern of this data set should not be allowed to hide two important characteristics: less developed countries that have high levels of exports in their GDPs are poised for entry into the next higher income groups; and less developed countries that are beginning to move into the international market—as evidenced by export share of GDP greater than 10 percent—are exhibiting potential for future economic growth.

– 59 –

Map 31 Exports of Primary Products

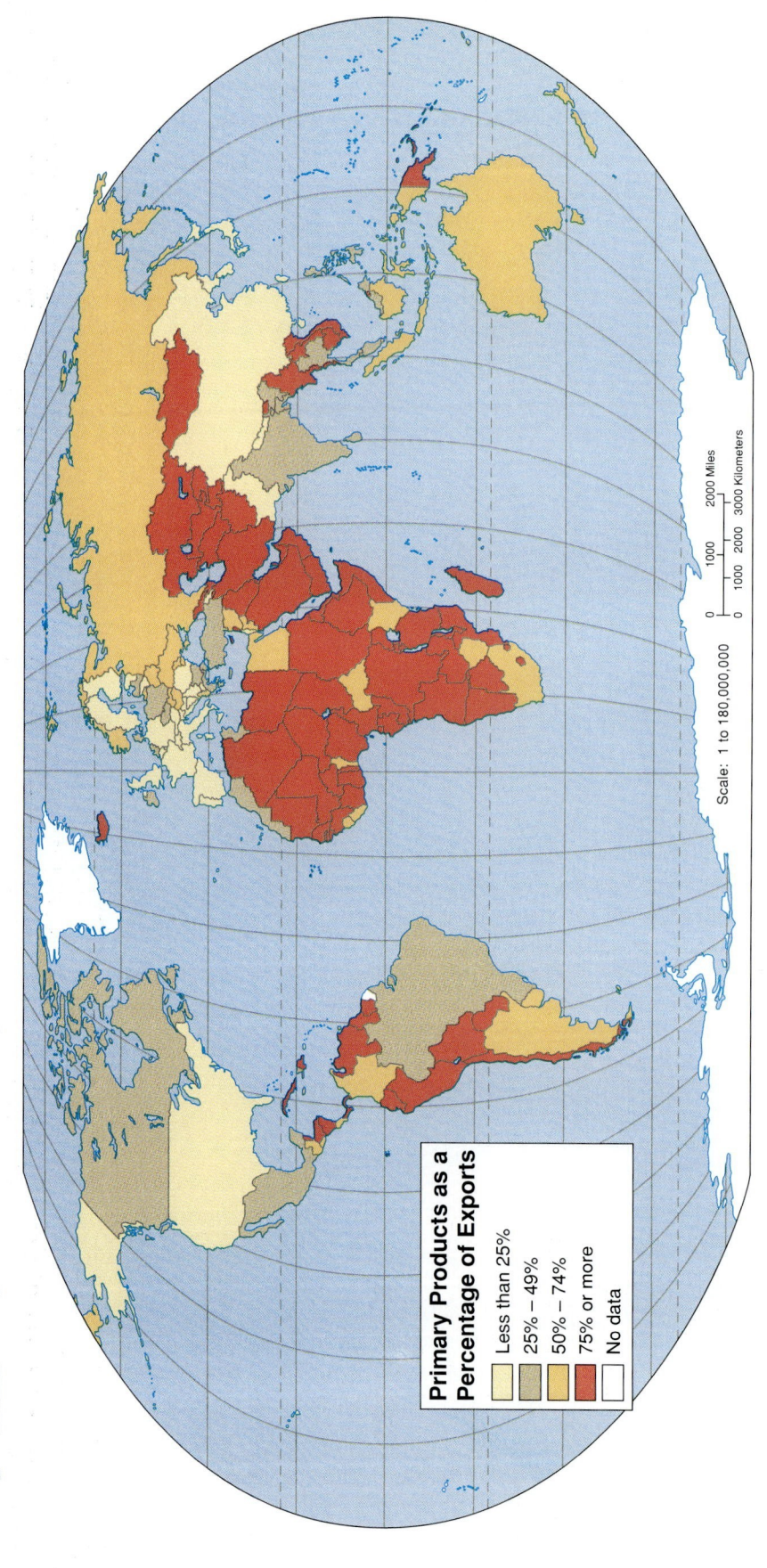

Primary Products as a Percentage of Exports
- Less than 25%
- 25% – 49%
- 50% – 74%
- 75% or more
- No data

Scale: 1 to 180,000,000

Primary products are those that require additional processing before they enter the consumer market: metallic ores that must be converted into metals and then into metal products such as automobiles or refrigerators; forest products such as timber that must be converted to lumber before they become suitable for construction purposes; and agricultural products that require further processing before being ready for human consumption. It is an axiom in international economics that the more a country relies on primary products for its export economy, the more vulnerable its economy is. Those countries with only primary products to export are hampered in their economic growth. A country dependent on only one or two products for export revenues is vulnerable to economic shifts, particularly to a changing market demand for its products. Imagine what would happen to the thriving economic status of the oil-exporting states of the Persian Gulf, for example, if an alternate source of cheap energy were found. A glance at this map, together with Map 22, shows that those countries with the lowest levels of economic development tend to be concentrated on primary products and, therefore, have economies that are especially vulnerable to economic instability.

– 60 –

Map 32 Dependence on Trade

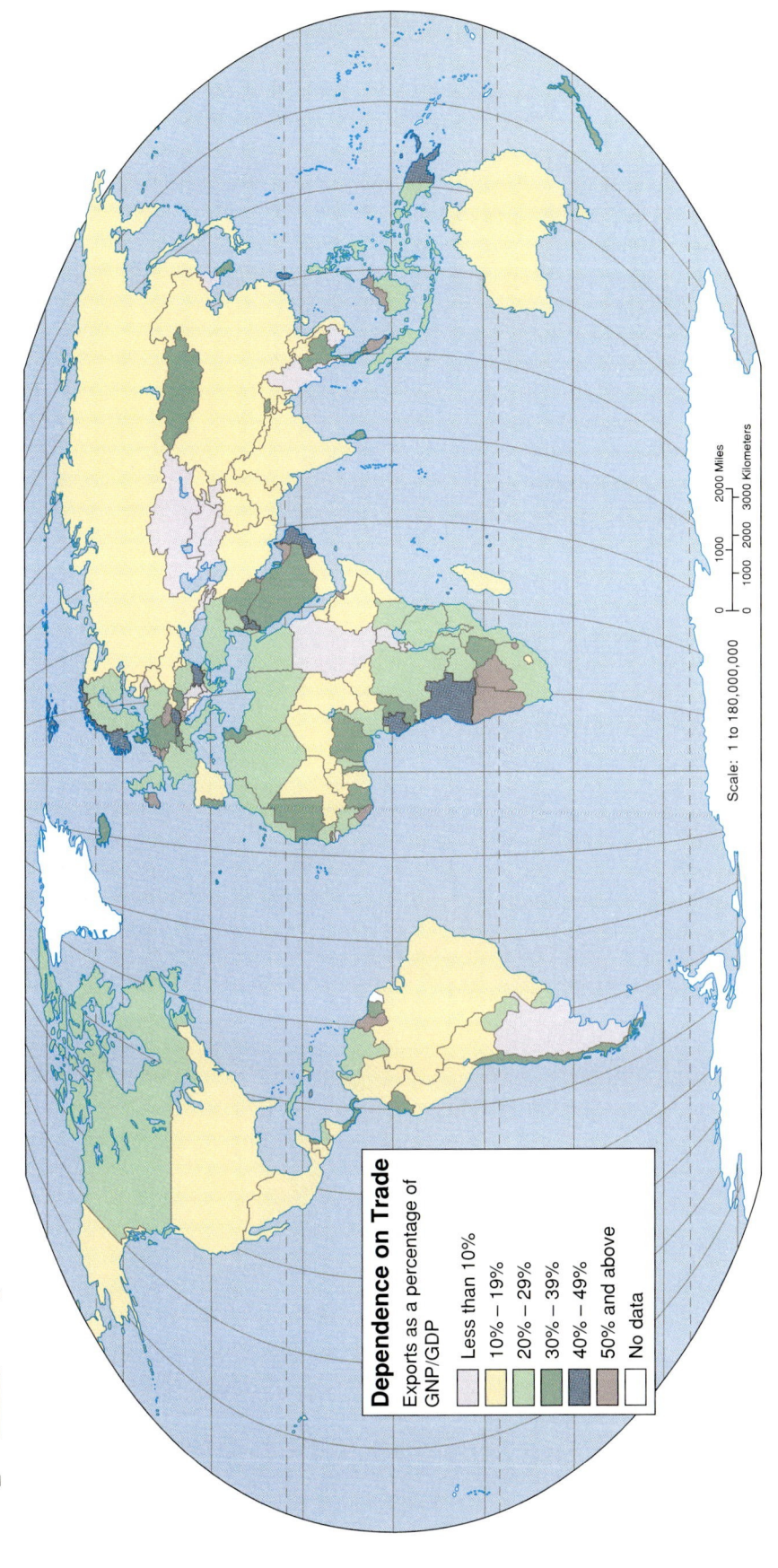

Dependence on Trade
Exports as a percentage of GNP/GDP

- Less than 10%
- 10% – 19%
- 20% – 29%
- 30% – 39%
- 40% – 49%
- 50% and above
- No data

Scale: 1 to 180,000,000

As the global economy becomes more and more a reality, the economic strength of virtually all countries is increasingly dependent upon trade. For many developing nations, with relatively abundant resources and limited industrial capacity, exports provide the primary base upon which their economies rest. Even countries like the United States, Japan, and Germany, with huge and diverse economies, depend on exports to generate a significant percentage of their employment and wealth. Without imports, many products that consumers want would be unavailable or more expensive; without exports, many jobs would be eliminated.

– 61 –

Map 33 Gross Domestic Product–Share in Investment

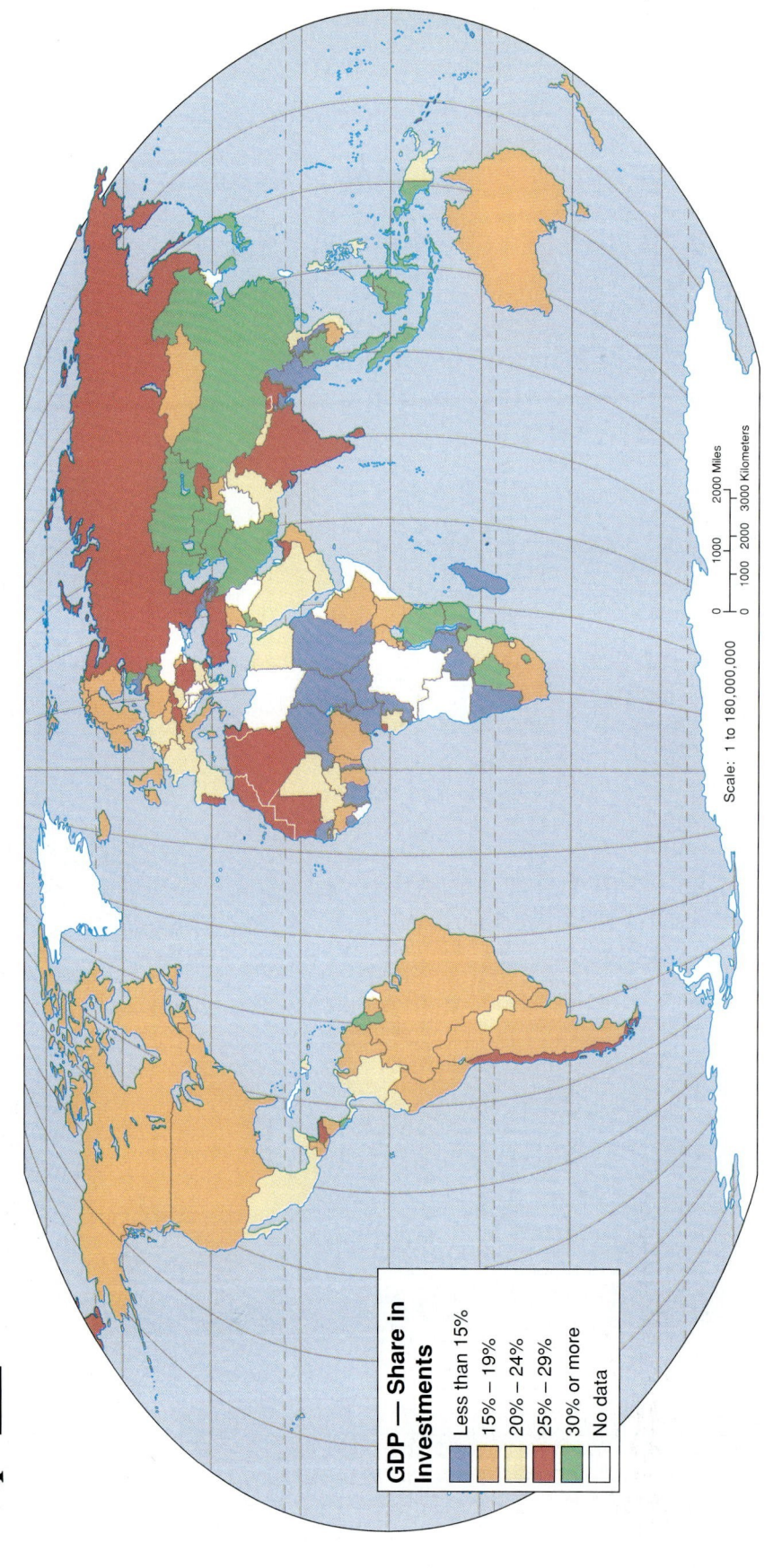

GDP — Share in Investments
- Less than 15%
- 15% – 19%
- 20% – 24%
- 25% – 29%
- 30% or more
- No data

Scale: 1 to 180,000,000

Few measures of economic development show as clear a geographic distribution as the share of Gross Domestic Product in investments, with North and South America, Europe, and Africa dominated by lower shares and Asia by higher shares. There are some cultural elements in this distribution, such as the historical tendency of Asian economies to be more investment-oriented. More importantly, the distribution reflects the recent tendency of low- and middle-income Asian countries to maintain or increase their level of investments, a trend that suggests favorable prospects for at least short-term growth in these economies. Africa still lags far behind because the very low levels of per capita GNP in most African countries provide little in the way of investment capital. Conversely, North American and European countries show low levels of investment share in GDP because their average GNP per capita is so high that even sizeable investments do not amount to large shares of GDP.

Map 34 Relative Wealth of Nations: Purchasing Power Parity

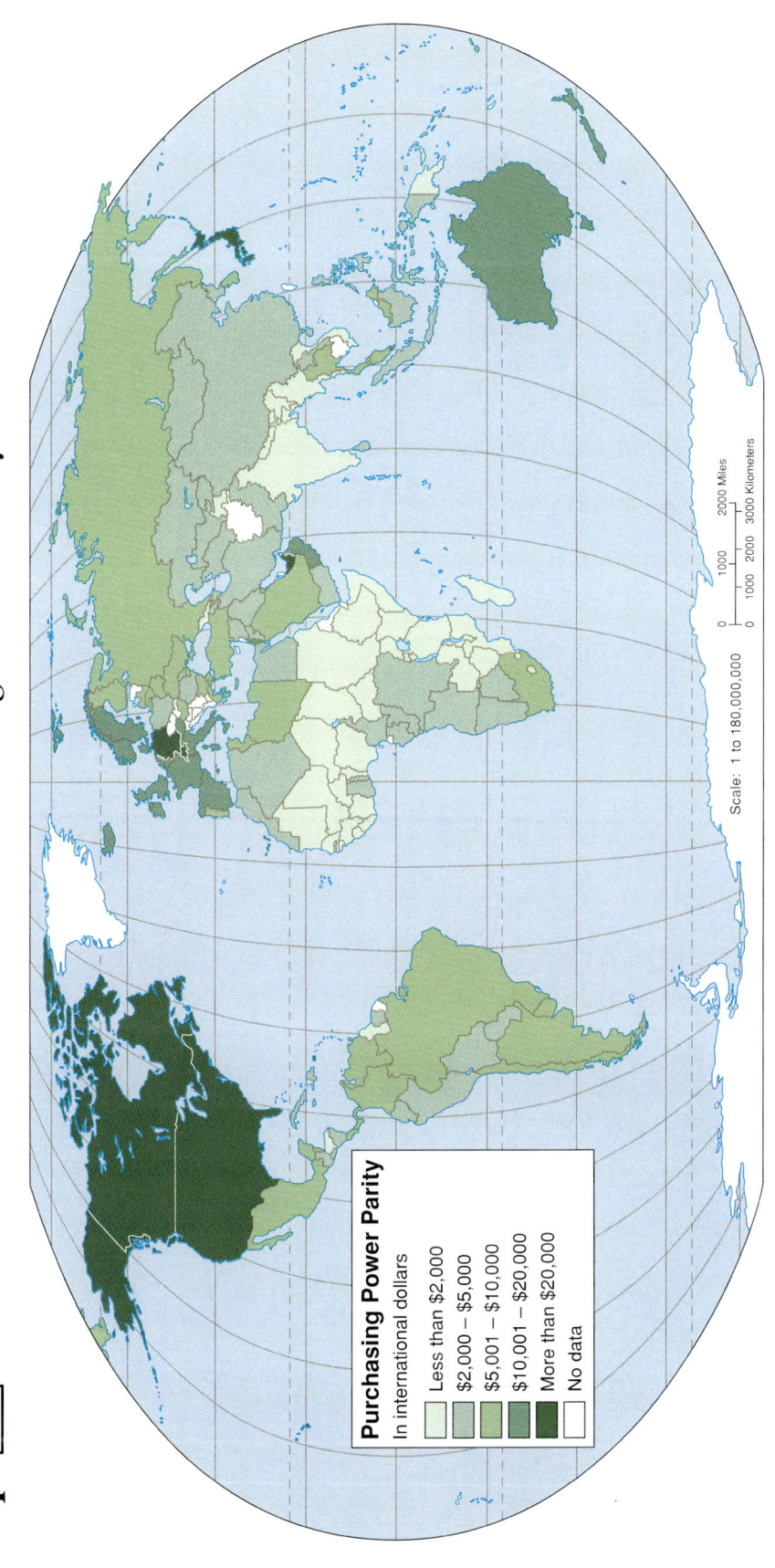

Purchasing Power Parity
In international dollars
- Less than $2,000
- $2,000 – $5,000
- $5,001 – $10,000
- $10,001 – $20,000
- More than $20,000
- No data

Scale: 1 to 180,000,000

Of all the economic measures that separate the "haves" from the "have-nots," perhaps per capita Purchasing Power Parity (PPP) is the most meaningful. While per capita figures can mask significant uneven distributions within a country, they are generally useful for demonstrating important differences between countries. Per capita GNP and GDP figures, and even per capita income, have the limitation of often not reflecting the true purchasing power of a country's currency at home. In order to get around this limitation, international economists seeking to compare national currencies developed the PPP measure, which shows the level of goods and services that holders of a country's money can acquire locally. By converting all currencies to the "international dollar," the World Bank and other organizations using PPP can now show more truly comparative values, since the new currency value shows the number of units of a country's currency required to buy the same quantity of goods and services as one U.S. dollar would buy in an average country. The use of PPP currency values can alter the perceptions about a country's true comparative position in the world economy. More than per capita income figures, PPP provides a valid measurement of the ability of a country's population to provide for itself the things that those of us in the developed nations take for granted: adequate food, shelter, clothing, education, and access to medical care. A glance at the map shows a clear-cut demarcation between temperate and tropical zones, with most of the countries with a PPP above $5,000 in the midlatitude zones and most of those with lower PPPs in the tropical and equatorial regions. Where exceptions to this pattern occur, they usually stem from a tremendous maldistribution of wealth among a country's population.

– 63 –

Map 35 Central Government Expenditures Per Capita

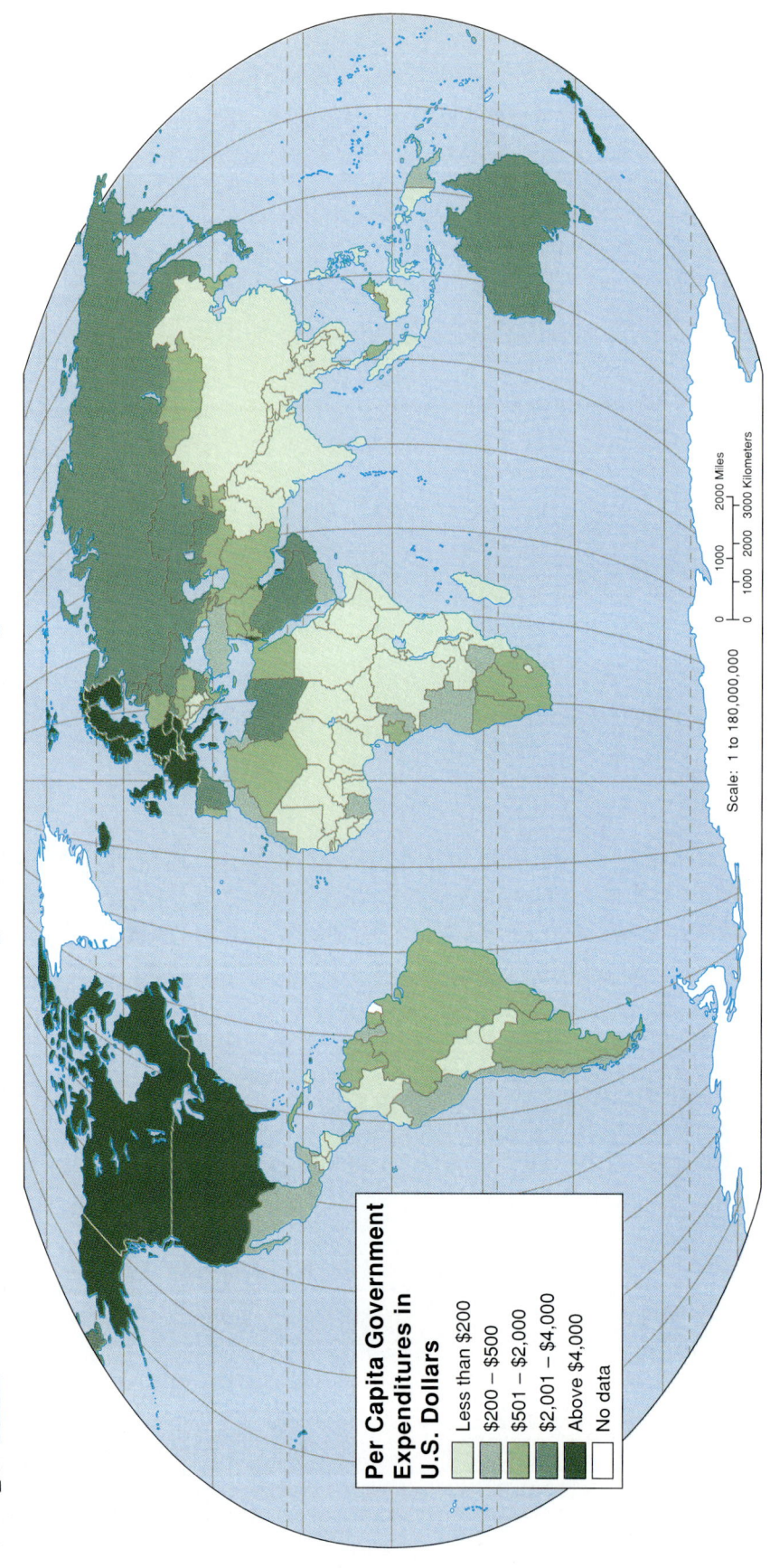

Per Capita Government Expenditures in U.S. Dollars
- Less than $200
- $200 – $500
- $501 – $2,000
- $2,001 – $4,000
- Above $4,000
- No data

Scale: 1 to 180,000,000

The amount of money that the central government of a country spends upon a variety of essential governmental functions is a measure of relative economic development, particularly when it is viewed on a per-person basis. These functions include such governmental responsibilities as agriculture, communications, culture, defense, education, fishing and hunting, health, housing, recreation, religion, social security, transportation, and welfare. Generally, the higher the level of economic development, the greater the per capita expenditures on these services. However, the data do mask some internal variations. For example, countries that tend to spend 20 percent or more of their central government expenditures on defense will often show up in the more developed category when, in fact, all that the figures really show is that a disproportionate amount of the money available to the government is devoted to purchasing armaments and maintaining a large standing military force. Thus, the fact that Libya spends $2,937 per capita—more than 10 times the average for Africa—does not suggest that the average Libyan is 10 times better off than the average Tanzanian. Nevertheless, this map—particularly when compared with Map 37, Energy Requirements Per Capita—does provide a reasonable approximation of economic development levels.

– 64 –

Map 36 Energy Production

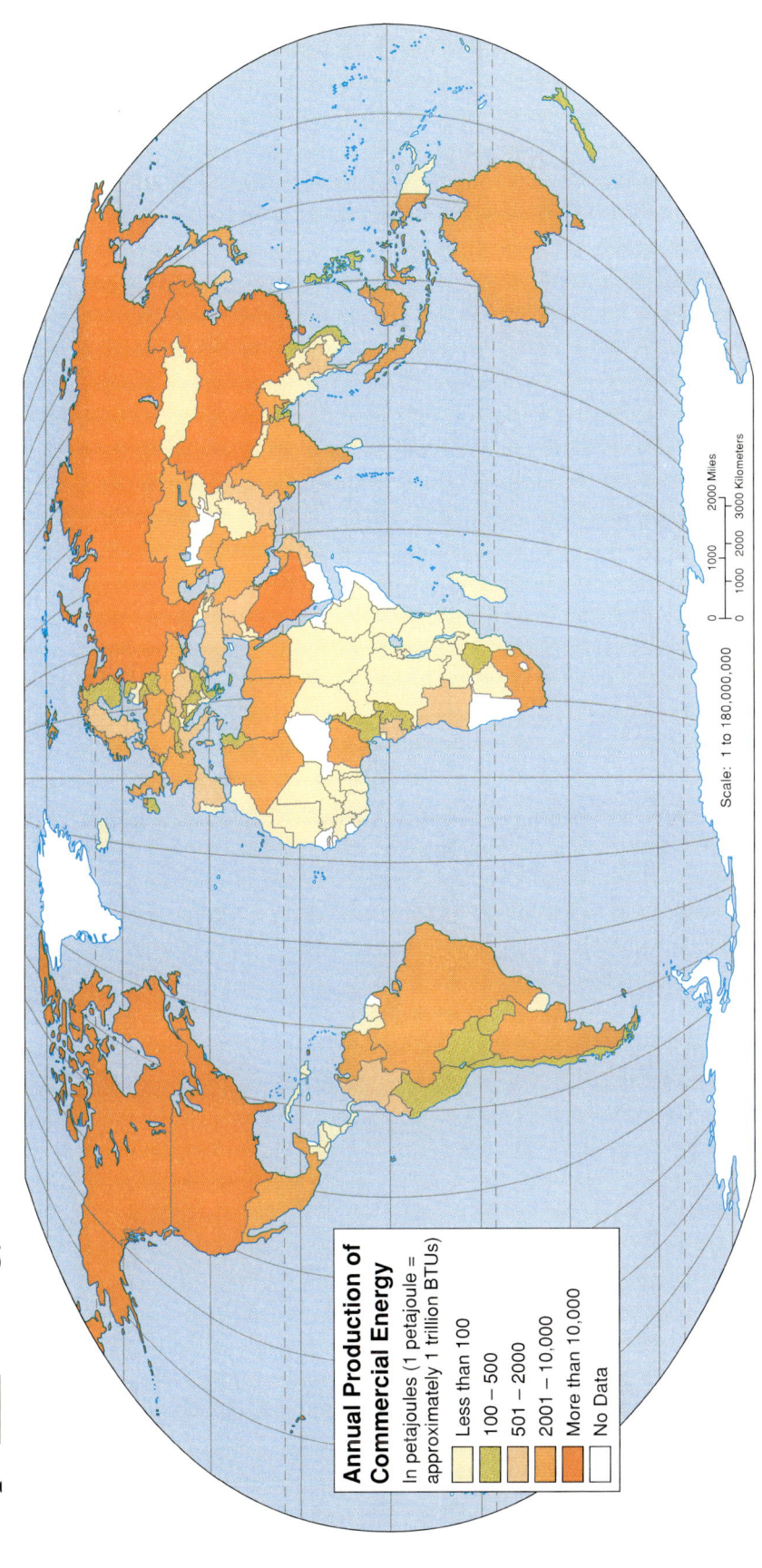

Annual Production of Commercial Energy

In petajoules (1 petajoule = approximately 1 trillion BTUs)

- Less than 100
- 100 – 500
- 501 – 2000
- 2001 – 10,000
- More than 10,000
- No Data

Scale: 1 to 180,000,000

It is one thing to have the significant energy resources that are possessed by a number of the world's countries; it is another thing to turn those resources into energy. The production of commercial energy in all its forms—solid fuels (primarily coal), liquid fuels (primarily petroleum), natural gas, geothermal, wind, solar, hydroelectric, and nuclear—is a good measure of economic potential, indicating the ability to produce sufficient quantities of energy either to meet domestic demands or to provide a healthy export commodity—or, in some instances, both. Commercial energy production is also a measure of the level of economic development, although a fairly subjective one. In general, the wealthier countries produce more energy from all sources (partly because they use more) than do the poorer countries. This general relationship has its exceptions, however. Countries such as Japan and many of the European nations rank among the world's wealthiest, but are energy-poor and produce relatively little of their own energy. They have the ability, however, to pay for it. On the other hand, countries such as those of the Persian Gulf or the oil-producing countries of Middle and South America may rank relatively low on the scale of economic development but rank high as producers of energy. The map does not show the enormous amounts of energy from noncommercial sources (traditional sources like firewood and animal dung) used by the world's poor, particularly in Middle and South America, Africa, South Asia, and East Asia. For a look at that kind of energy production, see Map 40. You may also be able to judge the variations between the production and use of commercial and noncommercial energy by comparing this map with the following one on energy consumption.

– 65 –

Map 37 Energy Requirements Per Capita

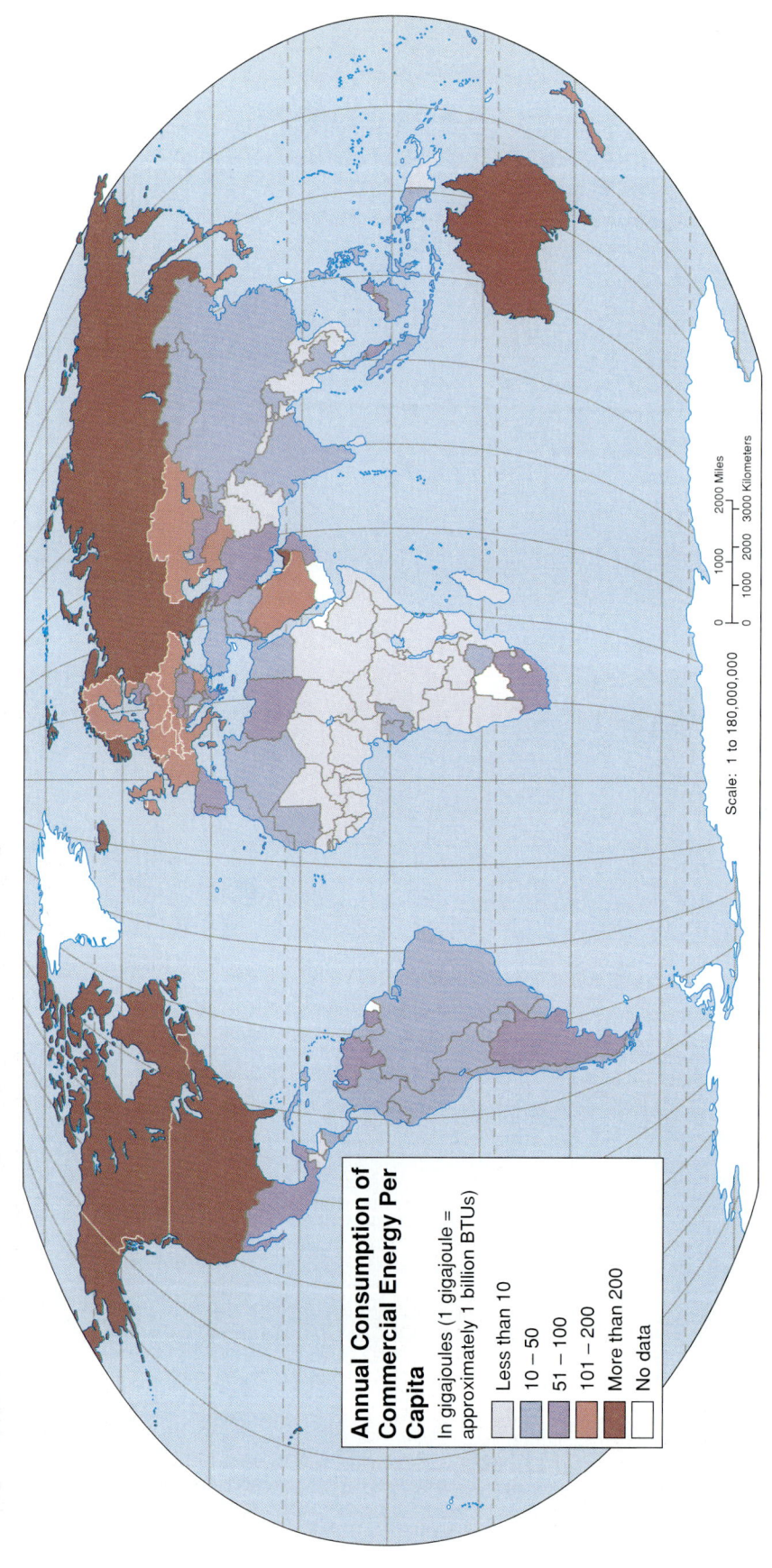

Annual Consumption of Commercial Energy Per Capita

In gigajoules (1 gigajoule = approximately 1 billion BTUs)

- Less than 10
- 10 – 50
- 51 – 100
- 101 – 200
- More than 200
- No data

Scale: 1 to 180,000,000

Of all the quantitative measures of economic well-being, energy consumption per capita may be the most expressive. Of the world's 25 countries defined as having high incomes, all consume at least 100 gigajoules of commercial energy (the equivalent of about 3.5 metric tons of coal) per person per year, with some, such as the United States and Canada, having consumption rates in the 300-gigajoule range (the equivalent of more than 10 metric tons of coal per person per year). With the exception of the oil-rich Persian Gulf states, where consumption figures include the costly "burning off" of excess energy in the form of natural gas, most of the highest-consuming countries are in the Northern Hemisphere, concentrated in North America and Western Europe. At the other end of the scale are the low-income countries where consumption rates are often less than 1 percent of those of the United States and other high consumers. These figures do not, of course, include the consumption of noncommercial energy—the traditional fuels of firewood, animal dung, and other organic matter—widely used in the less developed parts of the world.

— 66 —

Map 38 Production of Crucial Materials

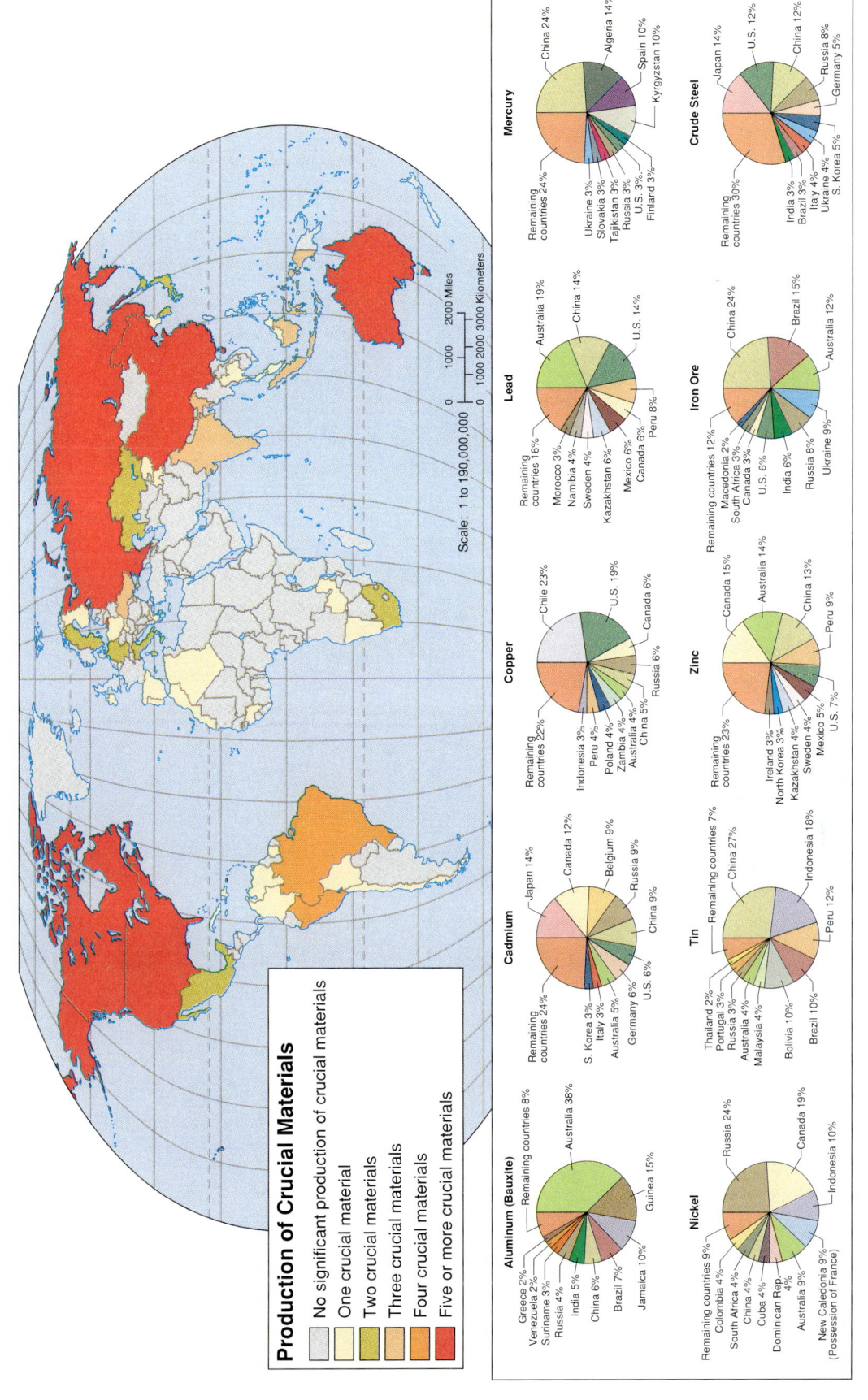

Production of Crucial Materials
- No significant production of crucial materials
- One crucial material
- Two crucial materials
- Three crucial materials
- Four crucial materials
- Five or more crucial materials

Scale: 1 to 190,000,000

Mercury
- Remaining countries 24%
- China 24%
- Algeria 14%
- Spain 10%
- Kyrgyzstan 10%
- Ukraine 3%
- Slovakia 3%
- Tajikistan 3%
- Russia 3%
- U.S. 3%
- Finland 3%

Crude Steel
- Remaining countries 30%
- Japan 14%
- U.S. 12%
- China 12%
- Russia 8%
- Germany 5%
- S. Korea 5%
- India 3%
- Brazil 3%
- Italy 3%
- Ukraine 4%

Lead
- Remaining countries 16%
- Australia 19%
- China 14%
- U.S. 14%
- Peru 8%
- Canada 6%
- Mexico 6%
- Kazakhstan 6%
- Sweden 4%
- Namibia 4%
- Morocco 3%

Iron Ore
- Remaining countries 12%
- China 24%
- Brazil 15%
- Australia 12%
- India 6%
- Russia 8%
- Ukraine 9%
- U.S. 6%
- Canada 3%
- South Africa 3%
- Macedonia 2%

Copper
- Remaining countries 22%
- Chile 23%
- U.S. 19%
- Canada 6%
- Russia 6%
- China 5%
- Australia 4%
- Zambia 4%
- Poland 4%
- Peru 4%
- Indonesia 3%

Zinc
- Remaining countries 23%
- Canada 15%
- Australia 14%
- China 13%
- Peru 9%
- U.S. 7%
- Mexico 5%
- Sweden 4%
- Kazakhstan 4%
- North Korea 3%
- Ireland 3%

Cadmium
- Remaining countries 24%
- Japan 14%
- Canada 12%
- Belgium 9%
- Russia 9%
- China 9%
- U.S. 6%
- Germany 4%
- Australia 5%
- Italy 3%
- S. Korea 3%

Tin
- Remaining countries 7%
- China 27%
- Indonesia 18%
- Peru 12%
- Brazil 10%
- Bolivia 10%
- Malaysia 4%
- Australia 4%
- Russia 3%
- Portugal 3%
- Thailand 2%

Aluminum (Bauxite)
- Remaining countries 8%
- Australia 38%
- Guinea 15%
- Jamaica 10%
- Brazil 7%
- China 6%
- India 5%
- Russia 4%
- Suriname 3%
- Venezuela 2%
- Greece 2%

Nickel
- Remaining countries 9%
- Russia 24%
- Canada 19%
- Indonesia 10%
- New Caledonia 9% (Possession of France)
- Australia 9%
- Dominican Rep. 4%
- Cuba 4%
- China 4%
- South Africa 4%
- Colombia 4%

The data on this map portray the production of the world's most important metals for the operation of a modern industrial economy. The sector graphs across the bottom of the map show the percentage of production of crucial materials by the 10 leading countries for each of 10 materials. For copper, lead, mercury, nickel, tin, and zinc, the annual production data reflect the metal content of the ore mined. Aluminum (or bauxite ore) and iron ore production are expressed in gross weight of ore mined. Cadmium production refers to the refined metal, and crude steel production to usable ingots, cast products, and liquid steel. By comparing this map with the following map, you will discover that some of the world's top consumer nations of these critical metals are also among the world's top producer nations.

Map 39 Consumption of Crucial Materials

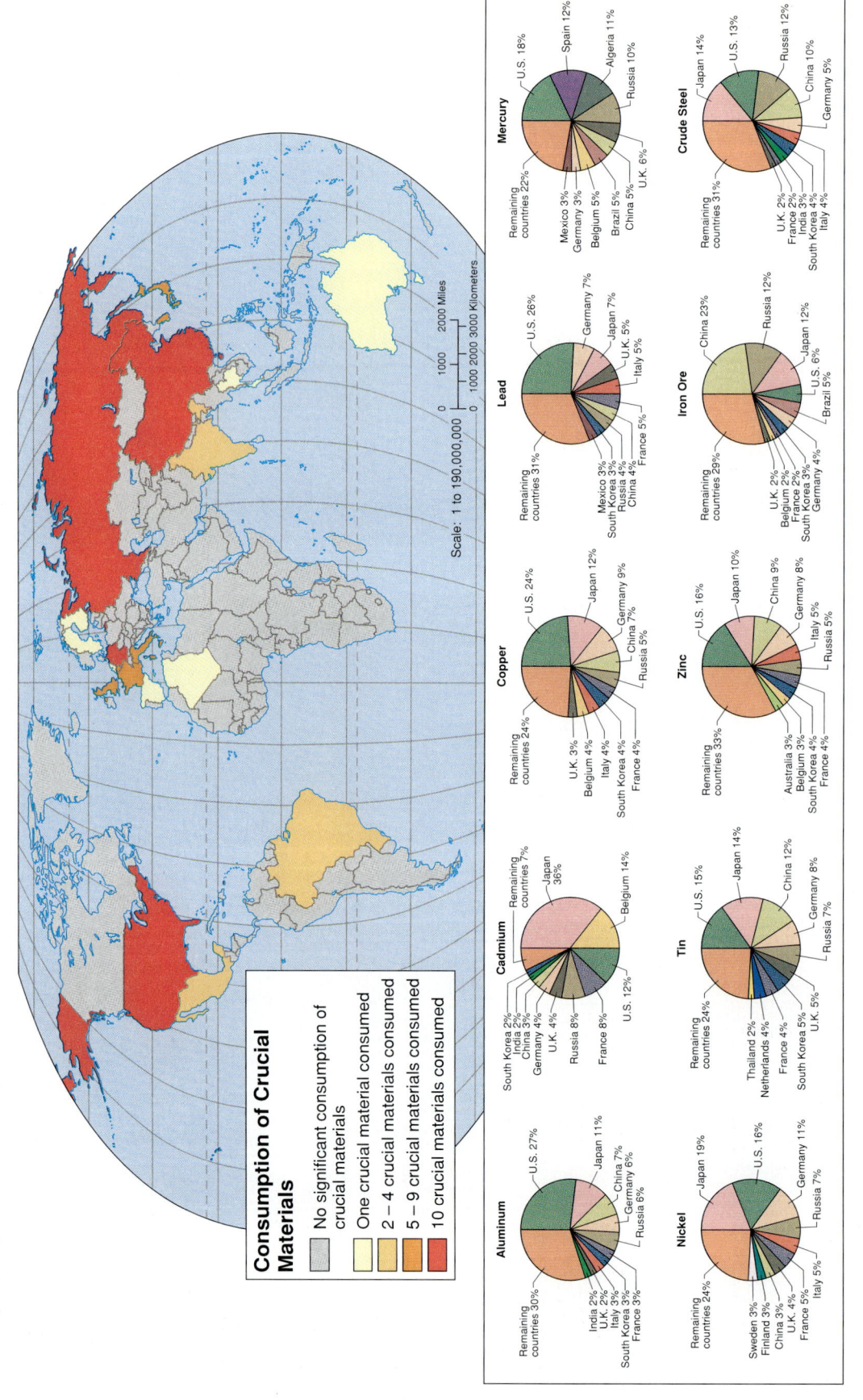

Consumption of Crucial Materials

- No significant consumption of crucial materials
- One crucial material consumed
- 2–4 crucial materials consumed
- 5–9 crucial materials consumed
- 10 crucial materials consumed

Scale: 1 to 190,000,000

Mercury: U.S. 18%, Spain 12%, Algeria 11%, Russia 10%, U.K. 6%, China 5%, Brazil 5%, Belgium 5%, Germany 3%, Mexico 3%, Remaining countries 22%

Crude Steel: Japan 14%, U.S. 13%, Russia 12%, China 10%, Germany 5%, Italy 4%, South Korea 4%, India 3%, France 2%, U.K. 2%, Remaining countries 31%

Lead: U.S. 26%, Germany 7%, Japan 7%, U.K. 5%, Italy 5%, France 5%, China 4%, South Korea 4%, Russia 4%, Mexico 3%, Remaining countries 31%

Iron Ore: China 23%, Russia 12%, Japan 12%, U.S. 6%, Brazil 5%, Germany 4%, South Korea 3%, France 2%, Belgium 2%, U.K. 2%, Remaining countries 29%

Copper: U.S. 24%, Japan 12%, Germany 9%, China 7%, Russia 5%, France 4%, South Korea 4%, Italy 4%, Belgium 4%, U.K. 3%, Remaining countries 24%

Zinc: U.S. 16%, Japan 10%, China 9%, Germany 8%, Italy 5%, Russia 5%, France 4%, South Korea 3%, Belgium 3%, Australia 3%, Remaining countries 33%

Cadmium: Japan 36%, Belgium 14%, U.S. 12%, France 8%, Russia 8%, U.K. 4%, Germany 4%, China 3%, India 2%, South Korea 2%, Remaining countries 7%

Tin: Japan 14%, U.S. 15%, China 12%, Germany 8%, Russia 7%, South Korea 5%, U.K. 5%, France 4%, Netherlands 4%, Thailand 2%, Remaining countries 24%

Aluminum: U.S. 27%, Japan 11%, China 7%, Germany 6%, Russia 6%, France 3%, South Korea 3%, Italy 3%, U.K. 2%, India 2%, Remaining countries 30%

Nickel: Japan 19%, U.S. 16%, Germany 11%, Russia 7%, Italy 5%, France 5%, U.K. 4%, China 3%, Finland 3%, Sweden 3%, Remaining countries 24%

Consumption data refers to the domestic use of refined metals (for example, the tons of steel used in the manufacture of automobiles). Some countries rank among the top in both production and consumption, and those that do are among the most highly developed nations. The United States, for example, ranks in the top 4 consumers for each metal; but the United States also ranks in the top 10 producer countries for 7 of the metals. Many countries that rank high as producers but not as consumers have "colonial dependency" economies, producing raw materials for an export market, often at the mercy of the marketplace. Jamaica and Suriname, for example, depend extremely heavily upon the sale of bauxite ore (crude aluminum). When the United States, Japan, or Russia cuts its use of aluminum, the economies of Jamaica and Suriname crash.

– 68 –

Table F
World Countries: Basic Economic Indicators

	GROSS NATIONAL PRODUCT 1993		PURCHASING POWER PARITY 1993		AVERAGE ANNUAL GROWTH RATE 1983–1993[1]			DISTRIBUTION OF GDP 1993			AVERAGE ANNUAL RATE OF INFLATION 1980–1993[1]
	TOTAL (millions $US)	PER CAPITA ($US)	TOTAL (millions $US)	PER CAPITA ($US)	GNP	GDP	PPP	AGRICULTURE	INDUSTRY	SERVICES	

AFRICA

Algeria	47,565	1,780	80,271	3,076	0.8	1.0	1.6	13	43	43	13.2
Angola	X	X	6,947	3,076	X	X	3.7	X	X	X	X
Benin	2,817	430	5,948	1,245	2.5	2.6	1.2	36	13	51	1.4
Botswana	3,909	2,790	4,202	3,406	9.6	8.8	6.4	6	47	47	12.3
Burkina Faso	2,932	300	6,183	651	3.2	3.2	3.3	44	20	37	3.3
Burundi	1,085	180	4,149	710	3.8	3.8	4.5	52	21	27	4.6
Cameroon	10,268	820	13,667	1,122	(1.9)	(2.2)	(0.4)	29	25	47	4.0
Central African Republic	1,262	400	1,950	634	0.4	0.8	0.8	50	14	36	4.2
Chad	1,262	210	2,955	504	4.2	4.2	5.3	44	22	35	0.7
Congo	2,321	950	6,012	2,538	0.7	0.5	1.0	11	35	53	(0.6)
Egypt	37,246	660	125,942	2,274	4.0	2.9	2.1	16	22	60	13.6
Equatorial Guinea	159	420	X	X	3.5	3.4	X	47	26	27	X
Eritrea	X	X	X	X	X	X	X	13	21	66	X
Ethiopia	X	100	17,245	405	X	X	0.1	60	10	29	X
Gabon	4,995	4,960	3,943	3,983	0.3	1.3	(0.2)	8	45	47	1.5
Gambia	365	350	940	1,019	4.3	3.2	5.1	28	15	58	16.2
Ghana	7,072	430	19,921	1,249	4.7	4.7	5.4	48	16	36	37.0
Guinea	3,153	500	3,694	604	4.2	3.7	3.5	24	31	45	X
Guinea-Bissau	247	240	832	827	5.2	5.0	2.4	45	19	36	58.7
Ivory Coast	8,389	630	16,882	1,315	(0.8)	(0.4)	(0.6)	37	24	39	1.5
Kenya	6,844	270	29,024	1,176	3.7	4.0	4.5	29	18	54	9.9
Lesotho	1,263	650	1,942	1,027	2.4	6.0	2.5	10	47	43	13.8
Liberia	1,313	560	2,272	1,001	(0.6)	(1.4)	(1.7)	X	X	X	X
Libya	26,840	6,125	36,531	9,649	(4.2)	(5.0)	X	X	X	X	X
Madagascar	3,048	220	10,148	757	1.3	1.4	0.5	34	14	52	16.1
Malawi	2,104	200	6,092	607	3.0	2.7	5.1	39	18	43	15.5
Mali	2,736	270	6,737	708	3.6	3.3	3.2	42	15	42	4.4
Mauritania	1,081	500	2,282	1,083	2.2	2.1	1.5	28	30	42	8.2
Mauritius	3,306	3,030	8,672	8,025	7.0	6.5	6.6	10	33	57	8.8
Morocco	26,983	1,040	70,474	2,777	3.3	3.6	4.1	14	32	53	6.6
Mozambique	1,359	90	13,369	898	3.1	4.5	(0.2)	33	12	55	42.3
Namibia	2,659	1,820	4,596	3,231	4.9	3.3	2.8	10	27	63	11.9
Niger	2,309	270	4,711	629	0.6	0.3	(0.5)	39	18	44	1.3
Nigeria	31,579	300	115,579	1,132	5.1	4.6	1.2	34	43	24	20.6
Rwanda	15,886	210	7,071	961	1.1	1.1	2.1	41	21	28	3.4
Senegal	5,927	750	10,601	1,411	2.6	2.4	2.4	20	19	61	4.9
Sierra Leone	670	150	3,984	914	(0.1)	1.4	0.0	38	16	46	61.1
Somalia	1,134	131	8,849	1,040	2.0	2.7	2.8	65	9	26	X
South Africa	118,184	2,980	150,608	3,885	1.2	1.0	1.4	5	39	56	14.7
Sudan	10,589	493	18,304	725	X	X	0.9	34	17	50	X
Swaziland	1,047	1,190	2,259	2,950	4.0	3.9	4.1	12	39	50	X

(Continued on next page)

	GROSS NATIONAL PRODUCT 1993		PURCHASING POWER PARITY 1993		AVERAGE ANNUAL GROWTH RATE 1983–1993[1]			DISTRIBUTION OF GDP 1993			AVERAGE ANNUAL RATE OF INFLATION 1980–1993[1]
	TOTAL (millions $US)	PER CAPITA ($US)	TOTAL (millions $US)	PER CAPITA ($US)	GNP	GDP	PPP	AGRICULTURE	INDUSTRY	SERVICES	
Tanzania	2,522	90	15,912	663	4.0	4.9	8.2	56	14	30	24.3
Togo	1,321	340	2,517	669	1.2	0.8	2.4	49	18	33	3.7
Tunisia	14,889	1,720	32,192	3,807	3.7	3.7	3.7	18	31	51	7.1
Uganda	3,245	180	11,504	654	3.9	3.8	(1.2)	53	12	35	X
Zaire	9,574	264	19,049	526	2.1	(0.5)	2.6	X	X	X	X
Zambia	3,396	380	7,360	877	1.4	1.3	1.0	34	36	40	58.9
Zimbabwe	5,584	520	15,419	1,479	2.9	2.9	2.3	15	36	48	14.4
NORTH AND MIDDLE AMERICA											
Belize	500	2,450	1,142	5,739	7.3	7.3	6.4	19	28	53	X
Canada	574,786	19,970	596,557	20,970	2.4	2.4	3.0	3	32	65	3.9
Costa Rica	7,031	2,150	14,403	4,522	5.0	4.4	4.2	15	26	59	22.1
Cuba	X	X	43,907	4,266	X	X	X	X	X	X	X
Dominican Republic	9,278	1,230	21,555	2,918	3.4	3.0	2.3	15	23	62	25.0
El Salvador	7,282	1,320	12,286	2,274	2.8	2.6	2.2	9	25	66	17.0
Guatemala	11,032	1,100	28,135	2,888	2.8	2.8	3.0	25	19	55	16.8
Haiti	3,158	477	6,794	1,069	(1.5)	(1.6)	0.3	39	16	46	X
Honduras	3,201	600	9,274	1,792	3.1	3.5	3.1	20	30	50	8.2
Jamaica	3,472	1,440	7,273	3,053	1.1	2.9	1.8	8	41	51	22.4
Mexico	324,997	3,610	692,795	7,867	2.4	2.0	3.6	8	28	63	57.9
Nicaragua	1,399	340	5,669	1,542	(4.1)	(2.6)	(3.6)	30	20	50	664.6
Panama	6,599	2,600	10,208	4,102	1.4	1.1	0.2	10	18	72	2.1
Trinidad and Tobago	4,895	3,380	12,679	10,145	(1.8)	(1.4)	(1.6)	3	43	55	4.8
United States	6,378,873	24,070	5,925,080	23,220	2.5	2.6	2.6	X	X	X	3.8
SOUTH AMERICA											
Argentina	243,877	7,220	191,699	5,921	1.8	1.4	(0.9)	6	31	63	374.3
Bolivia	5,369	760	14,288	2,066	3.0	2.4	1.5	X	X	X	187.1
Brazil	458,504	2,930	754,804	4,912	2.6	2.2	2.3	11	37	52	423.4
Chile	43,816	3,170	85,948	6,326	7.5	6.8	6.3	X	X	X	20.1
Colombia	49,955	1,400	148,368	4,254	3.9	4.0	4.0	16	35	50	24.9
Ecuador	13,176	1,200	36,685	3,420	3.1	2.7	1.9	12	38	50	40.4
Guyana	286	350	1,135	1,426	(0.3)	0.1	(1.7)	30	38	32	X
Paraguay	7,099	1,510	12,116	2,655	3.3	3.6	3.4	26	21	53	25.0
Peru	34,100	1,490	58,791	2,620	(0.6)	(0.5)	(0.7)	11	43	46	316.1
Suriname	489	1,180	1,102	2,787	0.8	0.9	(4.3)	22	24	54	X
Uruguay	12,061	3,830	21,087	6,736	3.9	3.0	3.3	9	27	64	66.7
Venezuela	59,393	2,840	172,419	8,449	3.1	3.1	2.9	5	42	53	23.9
ASIA											
Afghanistan	X	X	X	X	X	X	X	X	X	X	X
Armenia	2,462	660	16,477	4,750	(6.7)	(6.7)	0.0	48	30	22	26.9
Azerbaijan	5,390	730	27,016	4,257	(5.2)	(5.2)	X	22	52	26	28.2
Bangladesh	25,345	220	215,385	1,908	4.0	3.9	5.5	30	18	52	8.6
Bhutan	X	X	1,118	870	8.2	6.4	X	41	29	30	X
Cambodia	X	X	X	X	5.6	5.6	X	47	14	38	X

(Continued on next page)

	GROSS NATIONAL PRODUCT 1993		PURCHASING POWER PARITY 1993		AVERAGE ANNUAL GROWTH RATE 1983–1993[1]			DISTRIBUTION OF GDP 1993			AVERAGE ANNUAL RATE OF INFLATION 1980–1993[1]
	TOTAL (millions $US)	PER CAPITA ($US)	TOTAL (millions $US)	PER CAPITA ($US)	GNP	GDP	PPP	AGRICULTURE	INDUSTRY	SERVICES	
China	X	X	2,141,180	1,838	X	8.9	4.7	19	48	33	7.0
Georgia	3,159	580	24,388	4,495	(10.9)	(10.9)	X	58	22	20	40.7
India	269,460	300	1,437,124	1,633	5.0	5.1	5.2	31	27	41	8.7
Indonesia	138,492	740	478,799	2,601	5.9	5.8	5.0	19	39	42	8.5
Iran	134,174	2,159	258,602	4,161	1.6	1.7	1.7	21	36	43	17.1
Iraq	42,725	2,363	54,787	3,347	(14.9)	(14.9)	(9.0)	X	X	X	X
Israel	72,653	13,920	64,245	12,783	4.7	4.5	4.2	X	X	X	70.4
Japan	3,919,529	31,490	2,473,223	19,920	4.1	4.0	4.3	2	41	57	1.5
Jordan	4,881	1,190	13,241	4,039	0.1	1.2	1.1	8	26	66	X
Kazakhstan	26,445	1,560	82,590	4,929	(2.1)	(2.1)	X	29	42	30	35.2
Korea, North	X	X	60,990	3,067	X	X	X	X	X	X	X
Korea, South	338,044	7,660	415,320	9,565	9.0	8.7	9.6	7	43	50	6.3
Kuwait	34,120	19,360	17,557	8,561	(2.1)	2.0	(0.4)	0	55	45	X
Kyrgyzstan	3,902	850	14,959	3,372	0.6	0.6	X	43	35	22	28.6
Laos	1,289	280	7,592	1,753	4.7	4.7	3.9	51	18	31	X
Lebanon	X	X	X	X	X	X	X	X	X	X	X
Malaysia	59,808	3,140	133,586	7,191	6.7	6.5	6.2	X	X	X	2.2
Mongolia	904	390	5,319	2,443	X	X	3.7	21	46	33	13.8
Myanmar	X	X	31,582	772	(0.1)	(0.1)	1.0	63	9	28	16.5
Nepal	3,954	190	24,586	1,240	4.9	4.9	3.8	43	21	36	11.5
Oman	9,640	4,850	13,975	8,650	5.8	5.4	2.6	3	53	44	(2.3)
Pakistan	52,805	430	214,098	1,793	4.8	5.7	5.0	25	25	50	7.4
Philippines	55,080	850	137,734	2,172	2.6	2.1	2.4	22	33	45	13.6
Saudi Arabia	132,275	7,953	143,679	9,390	2.5	3.1	(1.6)	6	50	43	(2.1)
Singapore	55,380	19,850	46,213	16,736	6.9	6.9	6.1	0	37	63	2.5
Sri Lanka	10,738	600	49,170	2,783	3.9	3.8	2.8	25	26	50	11.1
Syria	15,582	1,219	63,326	4,955	0.8	1.9	2.6	30	23	48	X
Tajikistan	2,710	470	14,475	2,783	(3.0)	(3.0)	X	33	35	32	26.0
Thailand	122,515	2,110	286,553	5,018	8.9	8.8	7.8	10	39	51	4.3
Turkey	177,003	2,970	285,592	4,893	4.2	4.8	5.1	15	30	55	53.5
Turkmenistan	5,418	1,416	16,556	4,527	1.9	1.9	X	32	31	37	16.5
United Arab Emirates	38,727	21,430	25,381	15,784	1.6	(0.4)	0.5	2	57	40	X
Uzbekistan	21,204	970	72,104	3,334	1.4	1.4	X	23	36	41	24.5
Vietnam	12,125	170	39,838	665	X	6.6	X	29	28	42	X
Yemen	X	X	30,305	2,769	X	X	9.7	21	24	55	X
EUROPE											
Albania	1,152	340	X	X	X	(3.2)	X	40	13	47	5.6
Austria	184,829	23,150	132,669	16,969	2.7	2.6	2.7	2	35	62	3.6
Belarus	29,240	2,870	62,594	6,130	1.8	1.8	X	17	54	29	30.9
Belgium	217,537	21,650	181,190	18,091	2.6	2.4	2.8	2	30	68	4.0
Bosnia-Herzegovina	X	X	X	X	X	X	X	X	X	X	X
Bulgaria	10,112	1,140	60,299	6,774	(1.4)	(0.8)	2.5	13	38	49	15.9
Croatia	X	X	X	X	X	X	X	11	30	58	X
Czech Republic	27,902	2,710	X	X	(1.5)	(1.5)	1.3	6	40	54	X
Denmark	138,049	26,730	96,579	18,730	1.7	1.7	1.8	4	27	69	4.6

(Continued on next page)

	GROSS NATIONAL PRODUCT 1993		PURCHASING POWER PARITY 1993		AVERAGE ANNUAL GROWTH RATE 1983–1993[1]			DISTRIBUTION OF GDP 1993			AVERAGE ANNUAL RATE OF INFLATION 1980–1993[1]
	TOTAL (millions $US)	PER CAPITA ($US)	TOTAL (millions $US)	PER CAPITA ($US)	GNP	GDP	PPP	AGRICULTURE	INDUSTRY	SERVICES	
Estonia	4,780	3,080	9,906	6,326	(4.5)	(4.5)	X	8	29	63	29.8
Finland	97,624	19,300	78,627	15,619	1.1	1.4	1.7	5	31	64	5.8
France	1,292,556	22,490	1,043,232	18,232	2.2	2.2	2.5	3	29	69	5.1
Germany	1,901,131	23,560	1,315,229	20,197	X	X	X	1	38	61	2.8
Greece	76,599	7,390	91,259	8,877	1.8	1.8	2.4	18	32	50	17.3
Hungary	34,204	3,350	59,327	5,780	0.1	(1.2)	(1.1)	6	28	66	12.8
Iceland	6,570	24,950	4,253	16,324	2.3	2.2	2.4	12	28	60	X
Ireland	45,928	13,000	43,187	12,259	4.2	4.5	4.0	8	10	82	4.8
Italy	1,133,287	19,840	954,749	16,724	2.3	2.4	2.6	3	32	65	8.8
Latvia	5,248	2,010	18,246	6,891	(3.9)	(2.5)	X	15	32	53	23.8
Lithuania	4,900	1,320	18,637	5,025	(4.8)	(4.8)	X	21	41	38	35.2
Macedonia	1,702	820	X	X	X	X	X	X	X	X	X
Moldova	4,672	1,060	17,875	4,085	(3.8)	(3.8)	X	35	48	18	32.4
Netherlands	320,120	20,950	263,549	17,373	2.6	2.7	2.8	4	X	X	1.7
Norway	111,628	25,970	73,148	17,094	2.4	2.4	2.1	3	35	62	4.6
Poland	86,565	2,260	187,577	4,907	0.6	0.1	0.1	6	39	55	69.3
Portugal	89,848	9,130	95,107	9,638	4.2	3.6	5.6	6	38	56	16.4
Romania	25,948	1,140	48,388	2,130	(4.8)	(5.0)	1.9	21	40	40	22.4
Russia	347,896	2,340	1,223,573	8,320	(2.4)	(2.4)	X	9	51	39	35.4
Serbia-Montenegro (Yugoslavia)	X	X	X	X	2.9	2.9	(1.3)	X	X	X	X
Slovakia	10,360	1,950	X	X	(1.2)	(1.2)	X	7	44	49	X
Slovenia	12,571	6,490	X	X	X	X	X	6	36	58	X
Spain	536,547	13,590	511,794	12,986	3.4	3.4	4.2	5	35	61	8.4
Sweden	215,013	24,740	158,987	18,387	1.2	1.4	1.7	2	31	67	6.9
Switzerland	252,197	35,760	150,960	21,631	1.9	2.1	2.5	X	X	X	3.8
Ukraine	113,928	2,210	296,347	5,768	(1.2)	(1.2)	X	35	47	18	37.2
United Kingdom	1,045,994	18,060	941,413	16,302	2.2	2.3	2.7	2	33	65	5.6
OCEANIA											
Australia	307,967	17,500	321,126	18,500	2.7	2.8	2.8	3	29	67	6.1
Fiji	1,623	2,130	3,973	5,288	2.9	2.8	3.6	18	20	62	X
New Zealand	43,941	12,600	53,395	15,502	1.0	1.1	0.8	7	26	67	8.5
Papua New Guinea	4,644	1,130	7,923	1,972	3.2	3.7	1.4	26	43	31	4.8
Solomon Islands	262	740	789	2,639	5.0	5.4	7.7	X	X	X	X

[1] Negative numbers are shown in parentheses.

Sources: World Bank, Organization for Economic Co-Operation and Development; U.S. Central Intelligence Agency; *The World Almanac and Book of Facts 1996*; World Bank; *World Development Report 1995*; *World Resources 1996–97*.

Table G
World Countries: Agricultural Operations, 1993

NOTES: kg = kilogram; 1 kilogram = 2.205 lbs.
ha = hectare; 1 hectare = 2.471 acres

	CROPLAND (thousands of hectares) 1993	HECTARES PER CAPITA 1993	IRRIGATED LAND (% of cropland) 1991–1993	FERTILIZER USE (kg/ha cropland) 1993	TRACTORS AVERAGE NUMBER 1991–1993	TRACTORS % CHANGE SINCE 1981–1983[1]
WORLD	1,447,509	0.25	17	83	25,679,371	2
AFRICA	187,887	0.27	7	21	515,884	(32)
Algeria	7,850	0.29	7	21	91,333	88
Angola	3,500	0.34	2	2	10,297	X
Benin	1,880	0.37	0	9	137	22
Botswana	420	0.30	0	2	6,000	133
Burkina Faso	3,565	0.36	1	6	134	12
Burundi	1,360	0.23	1	3	166	53
Cameroon	7,040	0.56	0	3	503	(21)
Central African Republic	2,020	0.64	0	1	208	23
Chad	3,256	0.54	0	1	168	5
Congo	170	0.07	1	12	708	4
Egypt	2,800	0.05	100	357	60,333	33
Equatorial Guinea	230	0.61	0	0	100	2
Eritrea	1,280	0.36	1	0	850	X
Ethiopia	13,930	0.27	1	6	X	X
Gabon	460	0.37	1	1	1,493	14
Gambia	180	0.17	8	4	44	2
Ghana	4,320	0.26	0	1	4,083	12
Guinea	730	0.12	12	2	287	43
Guinea-Bissau	340	0.33	5	1	19	12
Ivory Coast	3,710	0.28	2	15	3,667	15
Kenya	4,520	0.17	1	27	14,000	73
Lesotho	320	0.16	1	19	1,850	22
Liberia	375	0.13	1	0	334	9
Libya	2,170	0.43	22	49	34,000	28
Madagascar	3,105	0.22	34	3	2,900	6
Malawi	1,700	0.16	1	51	1,417	11
Mali	2,503	0.25	0	10	840	1
Mauritania	208	0.10	24	22	336	8
Mauritius	106	0.10	16	245	367	9
Morocco	9,920	0.38	13	29	41,667	33
Mozambique	3,180	0.21	4	1	5,750	1
Namibia	662	0.45	1	0	3,133	16
Niger	3,605	0.42	2	0	179	38
Nigeria	32,385	0.31	3	16	11,867	25
Rwanda	1,170	0.15	0	2	90	7
Senegal	2,350	0.30	3	11	533	16
Sierra Leone	540	0.13	5	6	547	33
Somalia	1,020	0.11	18	0	2,137	16
South Africa	13,179	0.33	10	64	130,667	(24)
Sudan	12,975	0.49	15	5	10,484	6
Swaziland	191	0.24	35	61	4,433	27
Tanzania	3,500	0.12	4	14	6,633	(25)
Togo	2,430	0.63	0	4	370	42
Tunisia	4,952	0.58	8	22	26,833	2
Uganda	6,770	0.34	0	0	4,667	43
Zaire	7,900	0.19	0	1	2,427	14
Zambia	5,273	0.59	1	16	5,983	18
Zimbabwe	2,876	0.27	7	55	16,067	1
ASIA	468,661	0.14	34	118	5,565,425	38
Afghanistan	8,054	0.46	37	5	843	6
Armenia	X	X	X	X	14,615	9
Azerbaijan	2,000	0.27	58	27	35,667	(6)
Bangladesh	9,694	0.08	32	96	5,283	12

(Continued on next page)

	CROPLAND (thousands of hectares) 1993	HECTARES PER CAPITA 1993	IRRIGATED LAND (% of cropland) 1991–1993	FERTILIZER USE (kg/ha cropland) 1993	TRACTORS AVERAGE NUMBER 1991–1993	% CHANGE SINCE 1981–1983[1]
Bhutan	134	0.08	25	1	49	23
Cambodia	2,400	0.25	4	6	1,365	1
China	95,975	0.08	52	261	770,295	X
Georgia	1,000	0.18	44	54	22,500	(14)
India	169,650	0.19	28	73	1,131,395	124
Indonesia	30,987	0.16	15	85	34,660	234
Iran	18,150	0.28	51	52	117,333	19
Iraq	5,450	0.28	47	52	32,333	(4)
Israel	435	0.08	42	225	25,513	(6)
Japan	4,463	0.04	63	407	2,003,333	26
Jordan	405	0.08	16	34	5,800	21
Kazakhstan	34,800	2.05	6	14	203,333	26
Korea, North	2,000	0.09	73	315	74,667	21
Korea, South	2,055	0.05	65	474	64,644	747
Kuwait	5	0.00	40	200	107	357
Kyrgyzstan	1,420	0.31	67	20	25,133	(8)
Laos	805	0.17	15	4	887	24
Lebanon	306	0.11	28	118	3,000	0
Malaysia	4,880	0.25	7	212	12,433	38
Mongolia	1,401	0.60	6	4	11,667	9
Myanmar	10,087	0.23	10	19	10,667	18
Nepal	2,354	0.11	37	31	4,567	70
Oman	63	0.03	X	143	149	43
Pakistan	21,250	0.16	80	101	279,501	112
Philippines	9,190	0.14	17	61	11,333	22
Saudi Arabia	3,740	0.22	12	122	2,067	36
Singapore	1	0.00	X	5,600	65	29
Sri Lanka	1,900	0.11	29	111	32,733	25
Syria	5,775	0.42	15	65	69,198	88
Tajikistan	849	0.15	75	81	35,238	5
Thailand	20,800	0.36	21	54	74,870	192
Turkey	27,535	0.46	13	80	724,430	40
Turkmenistan	1,480	0.38	91	97	62,021	57
United Arab Emirates	39	0.02	13	710	184	11
Uzbekistan	4,500	0.21	92	150	178,333	4
Vietnam	6,700	0.09	28	136	36,667	29
Yemen	1,481	0.11	24	X	X	X
NORTH AND MIDDLE AMERICA	271,447	0.61	11	95	5,831,199	4
Belize	57	0.28	4	114	1,133	30
Canada	45,500	1.58	2	60	738,050	8
Costa Rica	530	0.16	23	208	6,833	12
Cuba	3,340	0.31	27	52	78,167	18
Dominican Republic	1,450	0.19	16	61	2,347	6
El Salvador	730	0.13	16	106	3,427	2
Guatemala	1,880	0.19	7	87	4,273	5
Haiti	910	0.13	8	5	228	20
Honduras	2,015	0.38	4	32	3,921	18
Jamaica	219	0.09	16	107	3,077	5
Mexico	24,730	0.27	24	71	172,000	14
Nicaragua	1,270	0.31	7	21	2,643	6
Panama	660	0.26	5	48	5,016	(6)
Trinidad and Tobago	122	0.10	18	51	2,643	6
United States	187,776	0.73	11	108	4,800,000	3
SOUTH AMERICA	102,767	0.33	34	118	5,565,425	38
Argentina	27,200	0.81	6	11	280,000	38
Bolivia	2,380	0.34	7	6	5,333	19
Brazil	48,955	0.31	6	85	733,333	19
Chile	4,257	0.31	30	58	39,727	(2)
Colombia	5,460	0.16	10	72	36,333	13

(Continued on next page)

	CROPLAND (thousands of hectares) 1993	HECTARES PER CAPITA 1993	IRRIGATED LAND (% of cropland) 1991–1993	FERTILIZER USE (kg/ha cropland) 1993	TRACTORS AVERAGE NUMBER 1991–1993	TRACTORS % CHANGE SINCE 1981–1983[1]
Ecuador	3,020	0.28	18	31	8,867	20
Guyana	496	0.61	26	24	3,627	3
Paraguay	2,270	0.48	3	14	16,293	70
Peru	3,430	0.15	37	44	16,500	6
Suriname	68	0.16	88	49	1,320	14
Uruguay	1,304	0.41	10	72	32,967	(2)
Venezuela	3,915	0.19	6	65	48,833	18
EUROPE	**136,005**	**0.19**	**12**	**116**	**9,791,242**	**6**
Albania	702	0.21	52	17	9,173	(11)
Austria	1,498	0.19	0	175	348,568	7
Belarus	6,248	0.61	2	136	125,167	(0)
Belgium	794	X	0	403	X	X
Bosnia-Herzegovina	940	0.25	0	11	205,930	X
Bulgaria	4,310	0.49	29	54	48,407	(17)
Croatia	1,313	0.29	0	172	4,200	(27)
Czech Republic	3,293	0.32	0	81	78,000	X
Denmark	2,542	0.49	17	191	157,069	(11)
Estonia	1,143	0.74	0	57	19,597	X
Finland	2,580	0.51	2	132	233,333	0
France	19,439	0.34	8	237	1,460,000	(0)
Germany	12,116	0.15	4	221	1,373,967	(16)
Greece	3,494	0.34	36	148	215,750	28
Hungary	4,973	0.49	4	40	41,380	(25)
Iceland	6	0.02	0	X	X	X
Ireland	923	0.26	9	769	167,333	10
Italy	11,860	0.21	23	148	1,439,022	23
Latvia	1,711	0.66	0	56	51,922	X
Lithuania	3,008	0.81	0	27	47,400	X
Macdeonia	663	0.31	12	18	49,864	X
Moldova	2,193	0.50	14	52	53,835	(0)
Netherlands	934	0.06	60	560	182,000	(0)
Norway	890	0.21	11	229	156,000	9
Poland	14,668	0.38	1	87	1,168,833	54
Portugal	3,160	0.32	20	75	131,158	30
Romania	9,941	0.43	31	39	142,114	(17)
Russia	133,900	0.91	4	29	1,283,333	(8)
Serbia-Montenegro (Yugoslavia)	X	X	X	X	X	X
Slovakia	1,613	0.30	2	50	33,663	X
Slovenia	301	0.16	1	249	69,761	X
Spain	19,656	0.50	17	93	765,769	29
Sweden	2,780	0.32	4	120	165,776	(12)
Switzerland	467	0.07	5	321	114,000	10
Ukraine	34,417	0.67	8	39	430,469	0
United Kingdom	6,127	0.11	2	338	500,000	(5)
OCEANIA	**51,500**	**1.86**	**5**	**41**	**401,469**	**(5)**
Australia	46,486	2.64	4	32	315,393	(3)
Fiji	260	0.34	0	56	7,051	43
New Zealand	3,800	1.09	7	154	75,667	(16)
Papua New Guinea	415	0.10	0	31	1,140	(8)
Solomon Islands	57	0.16	0	0	X	X

[1] Negative numbers are shown in parentheses.

Sources: Food and Agriculture Organization of the United Nations; United Nations Population Division; *World Resources 1996–97.*

Table H
World Countries: Energy Production, Consumption, and Requirements, 1993

	\multicolumn{4}{c	}{COMMERCIAL ENERGY}	\multicolumn{3}{c}{TRADITIONAL ENERGY}
	TOTAL PRODUCTION (in petajoules[1])	TOTAL CONSUMPTION (in petajoules)	PER CAPITA CONSUMPTION (in gigajoules)	IMPORTS AS % OF CONSUMPTION	TOTAL PRODUCTION (in petajoules)	PER CAPITA CONSUMPTION (in megajoules)	% OF TOTAL ENERGY CONSUMPTION
WORLD	337,518	325,295	59	X	19,925	3,594	6
AFRICA	21,308	8,805	13	(134)	4,815	6,991	35
Algeria	4,587	1,183	44	(274)	19	714	2
Angola	1,066	26	3	(3,835)	56	5,455	58
Benin	13	7	1	(71)	48	9,482	87
Botswana	X	X	X	X	13	9,420	100
Burkina Faso	X	8	1	100	85	8,652	91
Burundi	1	3	0	100	44	7,222	94
Cameroon	270	36	3	(639)	114	9,130	75
Central African Republic	0	3	1	133	34	10,694	92
Chad	X	1	0	200	35	5,900	97
Congo	365	24	10	(1,379)	22	8,945	48
Egypt	2,435	1,226	20	(84)	45	752	4
Equatorial Guinea	0	2	5	100	4	11,522	69
Eritrea	X	X	X	X	X	X	X
Ethiopia	7	45	1	93	414	7,984	90
Gabon	637	32	26	(1,859)	26	21,166	45
Gambia	X	3	3	100	9	8,579	75
Ghana	22	67	4	75	152	9,213	69
Guinea	1	15	2	100	35	5,594	70
Guinea-Bissau	X	3	3	100	4	4,012	58
Ivory Coast	18	109	8	119	103	7,723	49
Kenya	21	90	3	97	344	13,049	79
Lesotho	X	X	X	X	6	3,338	100
Liberia	1	5	2	100	48	17,045	91
Libya	3,054	457	91	(562)	5	1,037	1
Madagascar	1	15	1	93	76	5,483	84
Malawi	3	11	1	82	133	12,596	92
Mali	1	7	1	100	54	5,279	88
Mauritania	0	39	18	118	0	37	0
Mauritius	0	21	19	143	17	15,392	44
Morocco	21	297	11	108	14	529	4
Mozambique	1	14	1	114	147	9,758	91
Namibia	X	X	X	X	X	X	X
Niger	5	15	2	67	47	5,484	76
Nigeria	4,140	705	7	(481)	1,010	9,590	59
Rwanda	1	7	1	100	53	6,986	88
Senegal	X	38	5	126	49	6,257	57
Sierra Leone	X	6	1	233	30	6,903	83
Somalia	X	X	X	X	71	7,795	100
South Africa	4,146	3,578	79	(15)	131	3,314	4
Sudan	3	48	2	110	220	8,261	82
Swaziland	X	X	X	X	18	22,852	100
Tanzania	2	30	1	100	330	11,769	92
Togo	0	9	2	100	10	2,265	53
Tunisia	209	218	25	7	31	3,593	12
Uganda	3	16	1	81	137	6,870	90
Zaire	78	73	2	4	365	8,854	83
Zambia	38	51	5	33	130	14,536	72
Zimbabwe	160	208	19	25	70	6,513	25
NORTH AND MIDDLE AMERICA	87,427	97,154	220	11	1,825	4,130	2
Belize	X	4	20	110	4	18,789	49
Canada	13,195	9,198	319	(43)	67	2,325	1
Costa Rica	14	63	19	79	35	10,784	36
Cuba	43	369	34	95	205	18,848	36

(Continued on next page)

	COMMERCIAL ENERGY				TRADITIONAL ENERGY		
	TOTAL PRODUCTION (in petajoules[1])	TOTAL CONSUMPTION (in petajoules)	PER CAPITA CONSUMPTION (in gigajoules)	IMPORTS AS % OF CONSUMPTION	TOTAL PRODUCTION (in petajoules)	PER CAPITA CONSUMPTION (in megajoules)	% OF TOTAL ENERGY CONSUMPTION
Dominican Republic	6	148	20	96	25	3,360	15
El Salvador	21	72	13	74	39	7,050	35
Guatemala	22	72	7	89	104	10,335	59
Haiti	1	9	1	100	57	8,213	86
Honduras	8	43	8	81	58	10,897	57
Jamaica	0	104	43	100	5	2,493	5
Mexico	8,067	4,491	55	(57)	248	2,755	5
Nicaragua	20	52	13	67	39	9,450	43
Panama	8	61	24	89	16	6,366	21
Trinidad and Tobago	470	267	209	(78)	3	2,210	1
United States	65,547	81,751	317	21	916	3,553	1
SOUTH AMERICA	**15,355**	**10,095**	**33**	**(43)**	**2,748**	**8,888**	**21**
Argentina	2,411	2,019	60	(9)	116	3,421	5
Bolivia	164	86	12	(92)	19	2,723	18
Brazil	2,491	3,800	24	46	2,021	12,912	35
Chile	222	539	39	62	84	6,050	13
Colombia	1,812	829	24	(117)	235	6,927	22
Ecuador	784	245	22	(208)	74	6,757	23
Guyana	9	15	18	100	4	5,355	23
Paraguay	113	51	11	(131)	55	11,699	52
Peru	325	314	14	(3)	88	3,825	22
Suriname	15	24	58	75	1	2,959	5
Uruguay	26	77	24	69	28	8,948	27
Venezuela	6,990	2,083	100	(226)	22	1,046	1
ASIA	**113,332**	**95,679**	**28**	**(9)**	**9,009**	**2,690**	**9**
Afghanistan	9	22	1	64	51	2,863	70
Armenia	11	49	14	96	0	X	X
Azerbaijan	714	546	74	23	0	0	0
Bangladesh	217	313	3	33	277	2,401	47
Bhutan	6	2	1	(150)	12	7,345	85
Cambodia	0	7	1	100	54	5,560	88
China	31,359	29,679	25	(2)	2,018	1,687	6
Georgia	28	159	29	91	X	X	X
India	8,088	9,338	10	21	2,824	3,132	23
Indonesia	7,145	2,648	14	(125)	1,465	7,642	36
Iran	8,448	3,264	51	(164)	29	446	1
Iraq	1,476	933	48	(24)	1	53	0
Israel	1	505	96	118	0	24	0
Japan	3,466	17,505	141	87	10	78	0
Jordan	0	147	30	109	0	16	0
Kazakhstan	4,025	3,381	199	(16)	0	0	0
Korea, North	2,671	2,925	127	8	40	1,753	1
Korea, South	832	4,504	102	98	26	584	1
Kuwait	4,329	471	265	(798)	0	0	0
Kyrgyzstan	66	150	33	57	0	0	0
Laos	3	5	1	40	39	8,366	89
Lebanon	1	121	43	101	5	1,653	4
Malaysia	2,167	996	52	(114)	90	4,686	8
Mongolia	84	105	45	19	13	5,689	11
Myanmar	75	71	2	3	193	4,324	73
Nepal	3	19	1	84	206	9,882	92
Oman	1,720	162	81	(957)	0	0	0
Pakistan	766	1,135	9	36	296	2,228	21
Philippines	276	787	12	84	382	5,892	33
Saudi Arabia	19,171	2,933	171	0	0	0	0
Singapore	X	745	267	202	0	0	0
Sri Lanka	14	78	4	113	89	4,996	53
Syria	1,234	565	41	(105)	0	9	0
Tajikistan	70	258	45	75	0	0	0
Thailand	678	1,628	28	63	526	9,141	24
Turkey	779	1,979	33	67	96	1,606	5

(Continued on next page)

	COMMERCIAL ENERGY				TRADITIONAL ENERGY		
	TOTAL PRODUCTION (in petajoules[1])	TOTAL CONSUMPTION (in petajoules)	PER CAPITA CONSUMPTION (in gigajoules)	IMPORTS AS % OF CONSUMPTION	TOTAL PRODUCTION (in petajoules)	PER CAPITA CONSUMPTION (in megajoules)	% OF TOTAL ENERGY CONSUMPTION
Turkmenistan	2,430	555	142	(327)	X	X	X
United Arab Emirates	5,273	1,039	572	(364)	0	0	0
Uzbekistan	X	1,903	87	(13)	0	0	0
Vietnam	489	316	4	(48)	251	3,516	44
Yemen	X	X	X	X	X	X	X
EUROPE	**92,937**	**108,523**	**148**	**18**	**552**	**761**	**1**
Albania	44	43	13	37	15	4,485	26
Austria	263	966	123	75	30	3,766	3
Belarus	122	1,249	123	91	X	X	X
Belgium	470	1,976	197	90	6	557	0
Bosnia-Herzegovina	14	29	8	52	X	X	X
Bulgaria	376	965	109	70	13	1,448	1
Croatia	179	263	58	43	0	0	0
Czech Republic	1,439	1,659	161	18	0	0	0
Denmark	525	762	148	31	5	943	1
Estonia	121	214	138	45	0	0	0
Finland	324	1,014	200	63	30	5,892	3
France	4,746	9,153	159	53	101	1,757	1
Germany	6,178	13,724	170	57	0	0	0
Greece	352	989	95	74	13	1,274	1
Hungary	533	990	97	53	24	2,319	2
Iceland	25	54	205	54	0	0	0
Ireland	152	428	121	67	0	139	0
Italy	1,226	6,749	118	74	48	848	1
Latvia	14	187	72	84	X	X	X
Lithuania	138	368	99	62	X	X	X
Macedonia	86	139	66	45	0	0	0
Moldova	1	234	53	102	X	X	X
Netherlands	3,112	3,306	216	13	2	150	0
Norway	6,365	904	210	(588)	9	2,198	1
Poland	3,878	4,056	106	4	X	X	X
Portugal	35	603	61	106	6	573	1
Romania	1,345	1,762	7	30	19	841	1
Russia	43,550	30,042	203	(40)	0	0	0
Serbia-Montenegro (Yugloslavia)	347	381	36	17	X	X	X
Slovakia	186	672	126	74	0	0	0
Slovenia	86	194	100	56	X	X	X
Spain	1,204	3,359	85	78	18	466	1
Sweden	950	1,660	191	46	122	14,062	7
Switzerland	386	985	139	57	14	2,052	1
Ukraine	4,501	8,058	156	46	0	0	0
United Kingdom	9,663	90,518	164	1	4	72	0
OCEANIA	**7,159**	**4,595**	**166**	**(61)**	**185**	**6,693**	**4**
Australia	6,658	3,917	222	(77)	109	6,191	3
Fiji	1	11	15	109	12	15,606	52
New Zealand	496	565	162	15	0	140	0
Papua New Guinea	2	33	8	97	60	14,550	64
Solomon Islands	X	2	6	X	3	9,107	62

[1] One petajoule = 1×10^{15} joules = 947,800,000,000 BTUs = 163,400 UN standard barrels of oil.
One gigajoule = 1×10^9 joules = 947,800 BTUs = .16 UN standard barrels of oil.
One magajoule = 1,000,000 joules = 947.8 BTUs = .00016 UN standard barrels of oil.

Sources: United Nations Statistical Division; *World Development Report 1995; World Resources 1996–97.*

Table I
World Countries: Infrastructure

	ELECTRIC POWER		TELECOMMUNICATIONS		PAVED ROADS		WATER		RAILWAYS	
	PRODUCTION (kwh/person) 1992	SYSTEM LOSSES (% of total output) 1992	TELEPHONE MAINLINES (per 100 persons) 1992	FAULTS (per 100 mainlines per year) 1992	ROAD DENSITY (km per million persons) 1992	ROADS IN GOOD CONDITION (% of paved roads) 1988	POPULATION WITH ACCESS TO SAFE WATER (% of total) 1991	LOSSES (% of total water provision) 1986	RAIL TRAFFIC UNITS (per 1,000 $US GDP) 1992	DIESELS IN USE (% of diesel inventory) 1992
AFRICA										
Algeria	701	15	37	82	2,043	40	X	X	76	88
Angola	X	X	X	X	X	X	X	X	X	X
Benin	44	1	3	150	241	26	50	25	X	X
Botswana	X	X	27	55	1,977	94	X	X	X	X
Burkina Faso	X	X	2	2	158	24	67	X	X	X
Burundi	X	X	2	81	177	58	38	46	X	X
Cameroon	223	4	5	74	299	38	34	X	96	72
Central African Republic	X	X	2	X	135	30	12	X	X	X
Chad	X	X	1	152	56	X	X	X	X	X
Congo	181	0	7	54	509	50	X	X	140	31
Egypt	849	12	39	X	633	39	41	X	465	75
Equatorial Guinea	X	X	X	X	X	X	X	X	X	X
Eritrea	X	X	X	X	X	X	X	X	X	X
Ethiopia	25	3	3	74	77	48	18	46	X	60
Gabon	928	11	23	74	511	30	72	22	90	X
Gambia	X	X	14	120	772	X	77	X	X	X
Ghana	386	2	3	159	474	28	56	47	28	X
Guinea	X	X	2	X	229	27	33	X	X	X
Guinea-Bissau	X	X	6	4	X	X	25	X	X	X
Ivory Coast	144	4	7	80	357	75	83	16	32	44
Kenya	X	X	X	X	X	X	X	X	X	X
Lesotho	X	X	6	X	452	53	46	X	X	X
Liberia	X	X	X	X	X	X	X	X	X	X
Libya	X	X	X	X	X	X	X	X	X	X
Madagascar	X	X	3	78	433	56	53	X	X	X
Malawi	X	X	3	X	278	56	49	X	26	70
Mali	X	X	1	X	308	63	66	X	104	48
Mauritania	X	X	3	165	804	58	100	X	X	X
Mauritius	X	X	72	67	1,549	95	X	5	125	81
Morocco	383	3	25	84	179	20	22	X	X	X
Mozambique	24	24	3	10	343	12	X	X	X	X
Namibia	X	X	40	78	2,722	X	59	X	X	X
Niger	X	X	1	79	400	60	42	X	17	18
Nigeria	137	39	3	327	376	67	64	X	X	X
Rwanda	X	X	2	38	162	41	51	X	75	68
Senegal	99	9	8	36	542	28	43	X	X	X
Sierra Leone	X	X	3	17	295	62	X	X	X	X
Somalia	X	X	X	X	X	X	X	X	X	X
South Africa	4,329	7	89	X	1,394	X	X	X	804	82
Sudan	X	X	X	X	X	X	X	X	X	X
Swaziland	X	X	X	X	X	X	X	X	X	X

(Continued on next page)

	ELECTRIC POWER		TELECOMMUNICATIONS		PAVED ROADS		WATER		RAILWAYS	
	PRODUCTION (kwh/person) 1992	SYSTEM LOSSES (% of total output) 1992	TELEPHONE MAINLINES (per 100 persons) 1992	FAULTS (per 100 mainlines per year) 1992	ROAD DENSITY (km per million persons) 1992	ROADS IN GOOD CONDITION (% of paved roads) 1988	POPULATION WITH ACCESS TO SAFE WATER (% of total) 1991	LOSSES (% of total water provision) 1986	RAIL TRAFFIC UNITS (per 1,000 $US GDP) 1992	DIESELS IN USE (% of diesel inventory) 1992
Tanzania	66	12	3	X	142	25	52	X	X	50
Togo	X	X	4	22	470	40	71	X	X	X
Tunisia	731	6	45	113	2,080	55	67	30	119	57
Uganda	X	X	2	58	118	10	15	X	20	67
Zaire	X	X	X	X	X	X	X	X	X	X
Zambia	900	11	9	33	795	40	59	X	169	44
Zimbabwe	790	7	12	215	1,406	27	36	X	523	83
NORTH AND MIDDLE AMERICA										
Belize	X	X	X	X	X	X	X	X	X	X
Canada	18,309	7	592	X	11,541	X	100	X	325	91
Costa Rica	X	X	102	X	1,756	22	94	X	X	46
Cuba	X	X	X	X	X	X	X	X	X	X
Dominican Republic	X	X	66	133	364	52	62	X	X	X
El Salvador	X	X	31	X	323	7	41	X	X	X
Guatemala	290	15	22	58	320	X	60	X	X	X
Haiti	X	X	X	X	X	X	X	X	X	X
Honduras	X	X	21	40	443	50	72	X	X	X
Jamaica	897	20	70	84	1,881	10	78	31	73	75
Mexico	1,381	14	80	X	1,019	85	53	X	X	X
Nicaragua	X	X	14	X	414	X	83	20	X	X
Panama	1,167	24	97	10	1,332	36	96	X	X	X
Trinidad and Tobago	3,122	13	142	6	1,724	72	X	X	X	X
United States	12,900	8	565	X	14,455	X	X	X	344	90
SOUTH AMERICA										
Argentina	1,670	15	123	13	1,856	35	64	X	120	68
Bolivia	349	14	33	28	258	21	46	X	81	62
Brazil	1,580	15	71	43	929	30	96	30	61	62
Chile	1,646	11	94	82	808	42	86	X	42	57
Colombia	1,032	18	85	83	383	42	X	38	5	35
Ecuador	675	14	48	197	476	53	58	47	X	X
Guyana	X	X	X	X	X	X	X	X	X	X
Paraguay	6,693	0	28	X	X	X	33	X	X	X
Peru	587	11	27	47	347	24	58	X	16	X
Suriname	X	X	X	X	X	X	X	X	X	X
Uruguay	2,842	14	168	16	2,106	26	X	X	13	62
Venezuela	3,404	15	91	5	10,269	40	89	X	X	X
ASIA										
Afghanistan	X	X	X	X	X	X	X	X	X	X
Armenia	1,850	22	157	X	2,024	X	X	X	X	X
Azerbaijan	2,699	13	89	X	X	X	X	X	X	X
Bangladesh	79	32	2	X	59	15	78	47	37	74
Bhutan	X	X	X	X	X	X	X	X	X	X

(Continued on next page)

	ELECTRIC POWER		TELECOMMUNICATIONS		PAVED ROADS		WATER		RAILWAYS	
	PRODUCTION (kwh/person) 1992	SYSTEM LOSSES (% of total output) 1992	TELEPHONE MAINLINES (per 100 persons) 1992	FAULTS (per 100 mainlines per year) 1992	ROAD DENSITY (km per million persons) 1992	ROADS IN GOOD CONDITION (% of paved roads) 1988	POPULATION WITH ACCESS TO SAFE WATER (% of total) 1991	LOSSES (% of total water provision) 1986	RAIL TRAFFIC UNITS (per 1,000 $US GDP) 1992	DIESELS IN USE (% of diesel inventory) 1992
Cambodia	X	X	X	X	X	X	X	X	X	X
China	647	7	10	X	X	X	71	X	847	82
Georgia	1,210	23	105	43	X	X	X	X	X	X
India	373	23	8	218	893	20	75	29	488	90
Indonesia	233	17	8	49	160	30	42	X	27	75
Iran	1,101	12	50	X	X	X	89	X	61	39
Iraq	X	X	X	X	X	X	X	X	X	X
Israel	4,870	3	353	21	2,658	X	100	X	26	82
Japan	7,211	4	464	2	6,426	X	X	X	147	88
Jordan	1,120	13	71	89	1,767	X	99	41	74	76
Kazakhstan	4,826	9	88	X	6,747	X	X	X	X	54
Korea, North	X	X	X	X	X	X	X	X	X	X
Korea, South	2,996	5	357	13	1,090	70	78	X	146	88
Kuwait	8,924	9	245	30	X	X	100	X	X	X
Kyrgyzstan	2,636	10	75	30	X	X	X	X	X	X
Laos	X	X	2	12	516	X	28	X	X	X
Lebanon	X	X	X	X	X	X	X	X	X	X
Malaysia	1,612	9	112	78	X	X	78	29	30	76
Mongolia	X	X	30	43	X	X	66	X	X	58
Myanmar	61	35	2	X	210	40	33	X	X	75
Nepal	45	24	3	168	139	66	37	45	X	X
Oman	2,729	1	74	24	2,992	18	57	X	X	78
Pakistan	435	17	10	120	826	31	50	40	137	X
Philippines	419	13	10	10	242	X	81	53	X	X
Saudi Arabia	4,417	9	93	24	3,601	X	95	8	X	90
Singapore	6,353	5	415	11	993	X	100	X	X	X
Sri Lanka	200	17	8	X	536	10	60	X	65	X
Syria	X	X	X	X	X	X	X	X	X	X
Tajikistan	3,001	7	48	218	X	X	X	X	X	X
Thailand	1,000	10	31	32	841	50	72	48	75	72
Turkey	1,154	13	160	27	5,514	X	91	44	65	76
Turkmenistan	3,411	11	65	53	X	X	X	X	X	X
United Arab Emirates	9,917	6	321	X	2,706	X	100	X	X	X
Uzbekistan	2,390	10	67	X	X	X	X	X	X	X
Vietnam	139	24	2	X	X	X	50	X	X	60
Yemen	156	11	11	22	372	39	X	45	X	X
EUROPE										
Albania	1,002	13	13	28	414	X	100	X	X	78
Austria	6,554	6	440	35	13,954	X	100	X	213	83
Belarus	3,692	11	169	X	4,707	X	X	X	X	92
Belgium	7,215	5	425	8	12,909	X	100	X	120	83
Bosnia-Herzegovina	X	X	X	X	X	X	X	X	X	X
Bulgaria	4,000	14	275	48	3,986	X	100	X	297	78

(Continued on next page)

	ELECTRIC POWER		TELECOMMUNICATIONS		PAVED ROADS		WATER		RAILWAYS	
	PRODUCTION (kwh/person) 1992	SYSTEM LOSSES (% of total output) 1992	TELEPHONE MAINLINES (per 100 persons) 1992	FAULTS (per 100 mainlines per year) 1992	ROAD DENSITY (km per million persons) 1992	ROADS IN GOOD CONDITION (% of paved roads) 1988	POPULATION WITH ACCESS TO SAFE WATER (% of total) 1991	LOSSES (% of total water provision) 1986	RAIL TRAFFIC UNITS (per 1,000 $US GDP) 1992	DIESELS IN USE (% of diesel inventory) 1992
Croatia	X	X	X	X	X	X	X	X	X	X
Czech Republic	5,740	7	176	X	X	X	X	X	X	X
Denmark	5,983	7	581	X	X	X	100	X	X	X
Estonia	7,599	9	215	45	13,741	X	X	X	89	77
Finland	11,409	5	544	11	5,180	X	100	X	X	99
France	8,089	6	525	8	9,429	X	100	X	180	93
Germany	6,693	2	457	14	13,008	X	100	X	140	88
Greece	3,624	7	487	80	X	100	X	X	107	47
Hungary	3,080	10	125	60	10,341	X	100	X	37	78
Iceland	X	X	X	X	7,756	X	X	X	369	X
Ireland	4,545	9	314	38	X	X	100	X	X	60
Italy	3,963	7	410	17	24,468	X	100	X	54	79
Latvia	1,460	26	247	26	5,283	X	X	X	96	93
Lithuania	5,505	9	222	46	4,437	X	X	X	X	64
Macedonia	2,812	8	148	13	9,529	X	X	X	X	X
Moldova	2,562	11	117	45	2,310	X	X	X	X	X
Netherlands	5,089	4	487	4	2,832	X	100	X	90	85
Norway	27,501	8	529	16	6,078	X	100	X	X	X
Poland	3,473	11	103	X	14,698	X	100	X	610	67
Portugal	3,055	11	306	52	6,132	69	100	X	97	86
Romania	2,386	10	113	116	6,130	50	100	28	X	52
Russia	6,820	8	154	X	3,431	30	X	X	X	X
Serbia-Montenegro (Yugoslavia)	X	X	X	X	X	X	X	X	X	X
Slovakia	4,251	8	167	23	X	X	77	X	X	X
Slovenia	6,238	5	247	X	5,525	X	X	X	X	81
Spain	4,022	9	353	6	8,540	X	100	X	67	87
Sweden	16,913	6	682	10	11,747	X	100	X	201	88
Switzerland	8,471	6	606	21	10,299	X	100	X	X	X
Ukraine	4,900	9	145	49	3,085	X	X	X	X	60
United Kingdom	5,660	8	473	16	6,224	X	100	X	64	74
OCEANIA										
Australia	9,221	7	471	X	16,221	X	100	X	75	81
Fiji	X	X	X	X	X	X	X	X	X	X
New Zealand	9,086	8	449	X	15,725	X	97	X	64	90
Papua New Guinea	X	X	9	X	196	34	33	X	X	X
Solomon Islands	X	X	X	X	X	X	X	X	X	X

Source: Human Development Report 1995.

Part IV

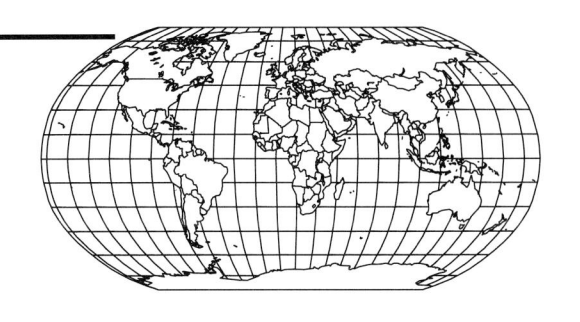

The Environmental Factor: World Environmental Conditions

Map 40 Reliance on Traditional Energy Sources: Per Capita Use of Biomass Energy

Map 41 Deforestation and Desertification

Map 42 Soil Degradation

Map 43 Annual Water Use

Map 44 Air and Water Quality

Map 45 Per Capita CO_2 Emissions

Table J Land Use

Table K Human-Induced Soil Destruction, 1945–1990

Table L Water Resources, 1995

Table M Greenhouse Gas Emissions, 1992

Map 40 Reliance on Traditional Energy Sources: Per Capita Use of Biomass Energy

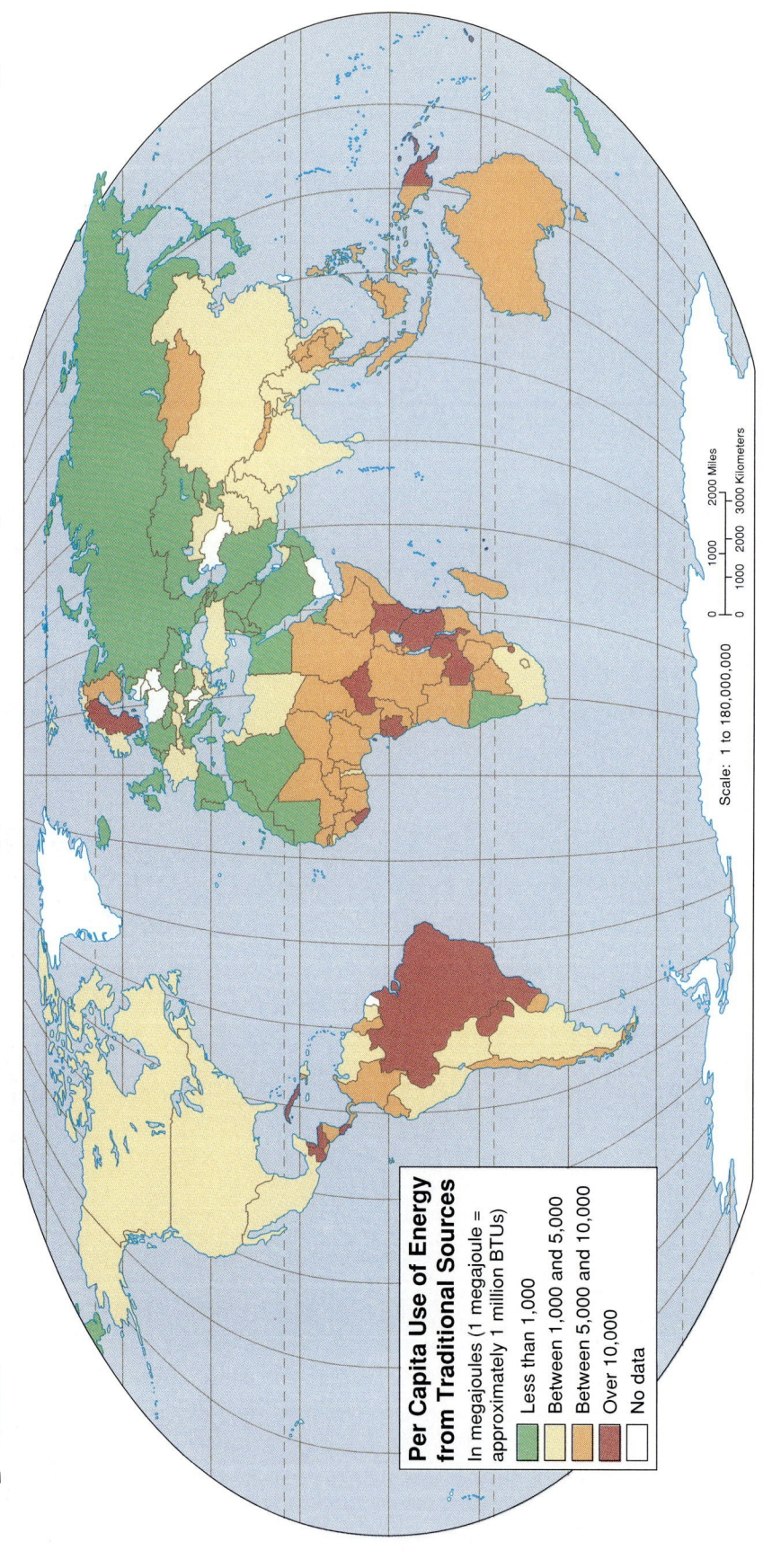

Per Capita Use of Energy from Traditional Sources

In megajoules (1 megajoule = approximately 1 million BTUs)

- Less than 1,000
- Between 1,000 and 5,000
- Between 5,000 and 10,000
- Over 10,000
- No data

Scale: 1 to 180,000,000

For much of the world's population, the primary energy source for cooking, heating, and other purposes is the same now as it has been for thousands of years: biomass, for the most part wood and animal dung. While the use of traditional fuels would seem to be ecologically sound, since the energy sources are renewable, using these fuels involves significant actual or potential environmental hazards. Increasing use of wood to meet the demands of a growing population means that firewood gatherers are forced to take down green, healthy trees rather than relying on downed branches. And where animal dung is used for burning, as it is throughout many of the world's drier areas, its potential to serve as a fertilizer to replenish poor soils is lost. As much as anything else, the reliance on traditional energy sources separates the "have-nots" from the "haves."

– 84 –

Map 41 Deforestation and Desertification

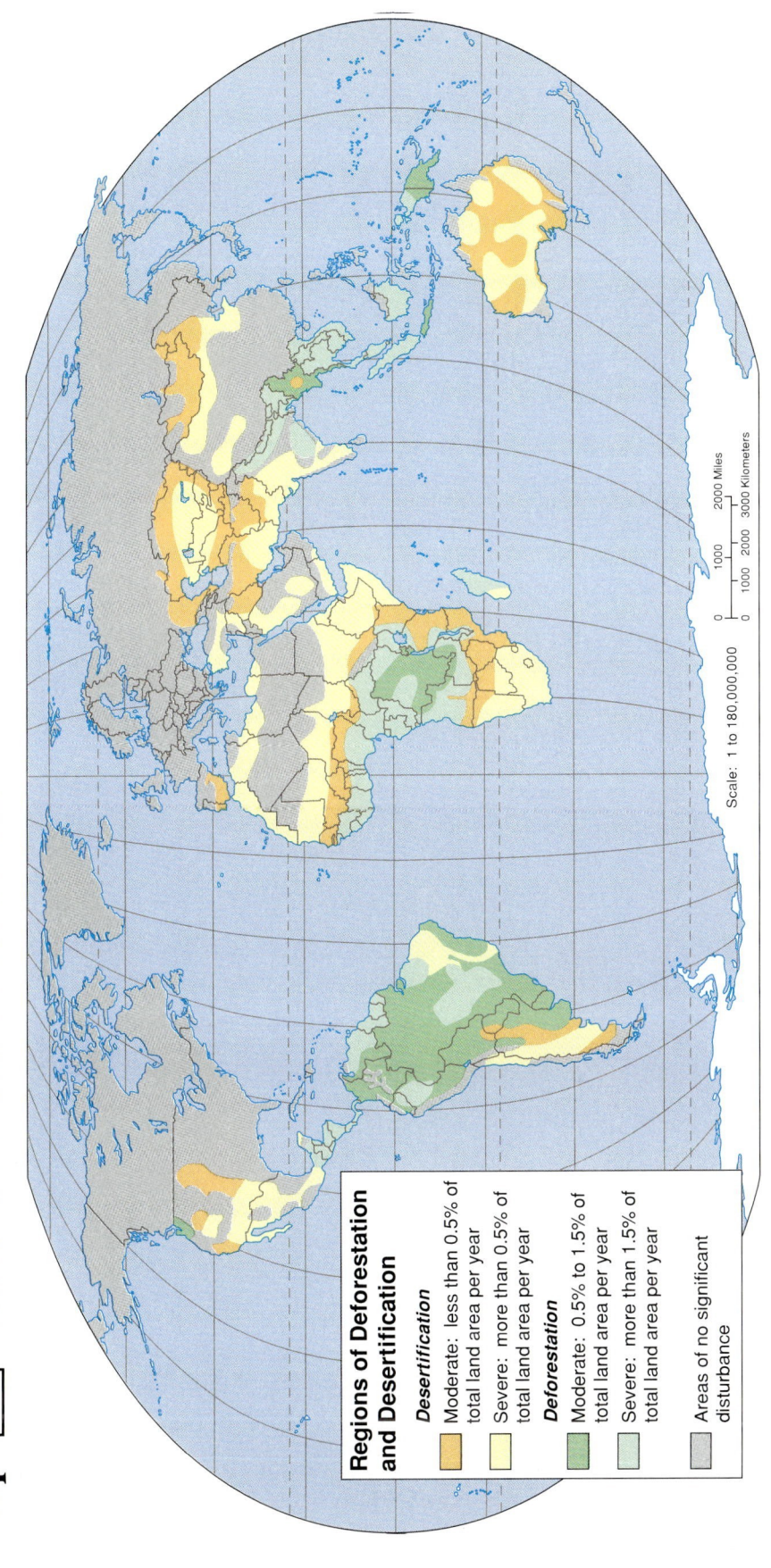

Regions of Deforestation and Desertification

Desertification
- Moderate: less than 0.5% of total land area per year
- Severe: more than 0.5% of total land area per year

Deforestation
- Moderate: 0.5% to 1.5% of total land area per year
- Severe: more than 1.5% of total land area per year
- Areas of no significant disturbance

Scale: 1 to 180,000,000

While those of us in the developed countries of the world tend to think of environmental deterioration as the consequence of our heavily industrialized economies, in fact the worst examples of current environmental degradation are to be found within the world's less developed regions. There, high population growth rates and economies limited primarily to farming have forced the increasing use of more marginal (less suited to cultivation) land. In the world's grassland and arid environments, which occupy approximately 40 percent of the world's total land area, increasing pressures of cultivation are turning vulnerable areas into deserts incapable of sustaining agricultural productivity. In the world's forested regions, particularly in the tropical forests of Africa, Asia, and Middle and South America, a similar process is occurring: increasing pressure for more farmland is creating a process of deforestation or forest clearing that destroys the soil, reduces the biological diversity of the forest regions, and—ultimately—may have the capacity to alter the global climate by contributing to an increase in carbon dioxide in the atmosphere. This increases the heat trapped in the atmosphere and enhances the greenhouse effect.

– 85 –

Map 42 Soil Degradation

Global Soil Degradation
- Areas of serious concern
- Areas of moderate concern
- Stable or nonvegetated areas

Scale: 1 to 180,000,000

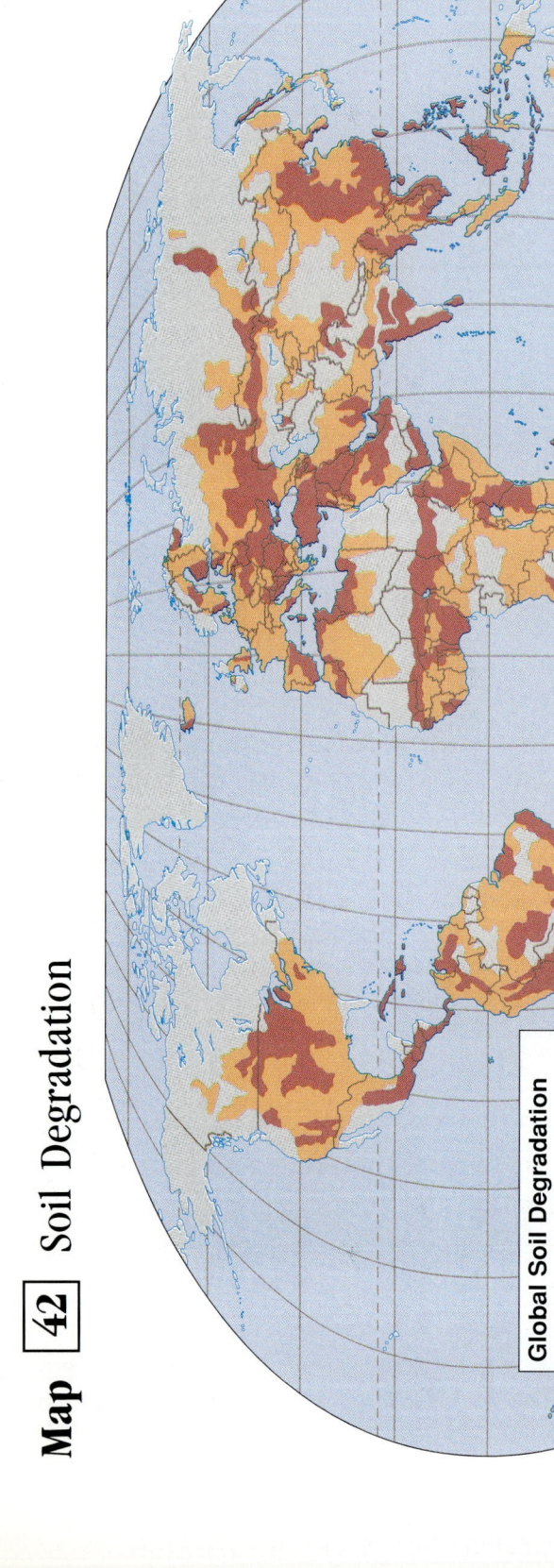

Recent research has shown that more than 3 billion acres of the world's surface suffer from serious soil degradation, with more than 22 million acres so severely eroded or poisoned with chemicals that they can no longer support productive crop agriculture. Most of this soil damage has been caused by poor farming practices, overgrazing of domestic livestock, and deforestation. These activities strip away the protective cover of natural vegetation—forests and grasslands—allowing wind and water erosion to remove the topsoil that contains the necessary nutrients and soil microbes for plant growth. But millions of acres of topsoil have been degraded by chemicals as well. In some instances these chemicals are the result of overapplication of fertilizers, herbicides, pesticides, and other agricultural chemicals or the consequence of the salt accumulation that accompanies irrigation agriculture. In other instances, chemical deposition from industrial and urban wastes and from acid precipitation have poisoned millions of acres of soil. As the map shows, soil erosion and pollution are not just problems in developing countries with high population densities and increasing use of marginal lands. They also afflict the more highly developed regions of mechanized, industrial agriculture. While many methods for preventing or reducing soil degradation exist, they are not used as often as they should be. The reasons are ignorance, cost, or perceived economic inefficiency.

– 86 –

Map 43 Annual Water Use

Annual Withdrawals of Water
In cubic meters per person

- Less than 100
- 100 – 500
- 501 – 1,000
- 1,001 – 2,000
- More than 2,000
- No data

Scale: 1 to 180,000,000

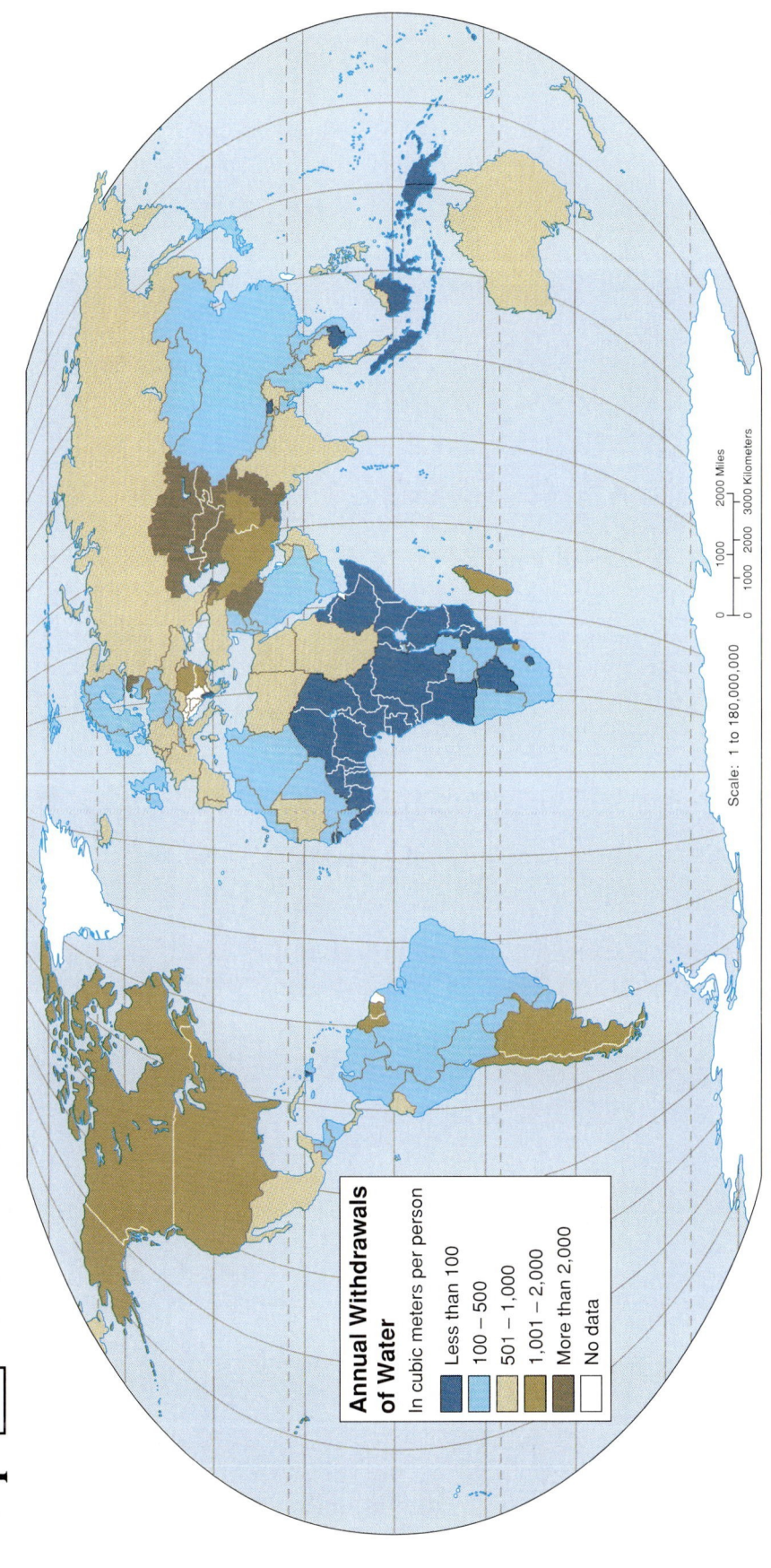

For thousands of years, human societies have understood water resources to be renewable, replenished by winter snows or spring rains. In fact, as long as water use is confined to the withdrawal of water from surface sources (lakes and streams), then that understanding is relatively accurate. In the modern world, however, the bulk of the water that is used for most purposes—industrial, agricultural, and domestic—is groundwater. Groundwater is water that has accumulated in a zone of varying depth and thickness below the surface. The quantity of groundwater is often the result of hundreds or thousands of years of accumulation from precipitation and, in some areas, melting glaciers. When humans begin to use or "withdraw" this water faster than it is replenished or "recharged," the consequence is a reduction of groundwater resources. Water is capable of being mined, just like petroleum; and just like petroleum, it can become depleted. The overuse of water (in which withdrawal exceeds recharge) is a result of economic expansion that fails to take into account the environmental costs of doing business.

– 87 –

Map 44 Air and Water Quality

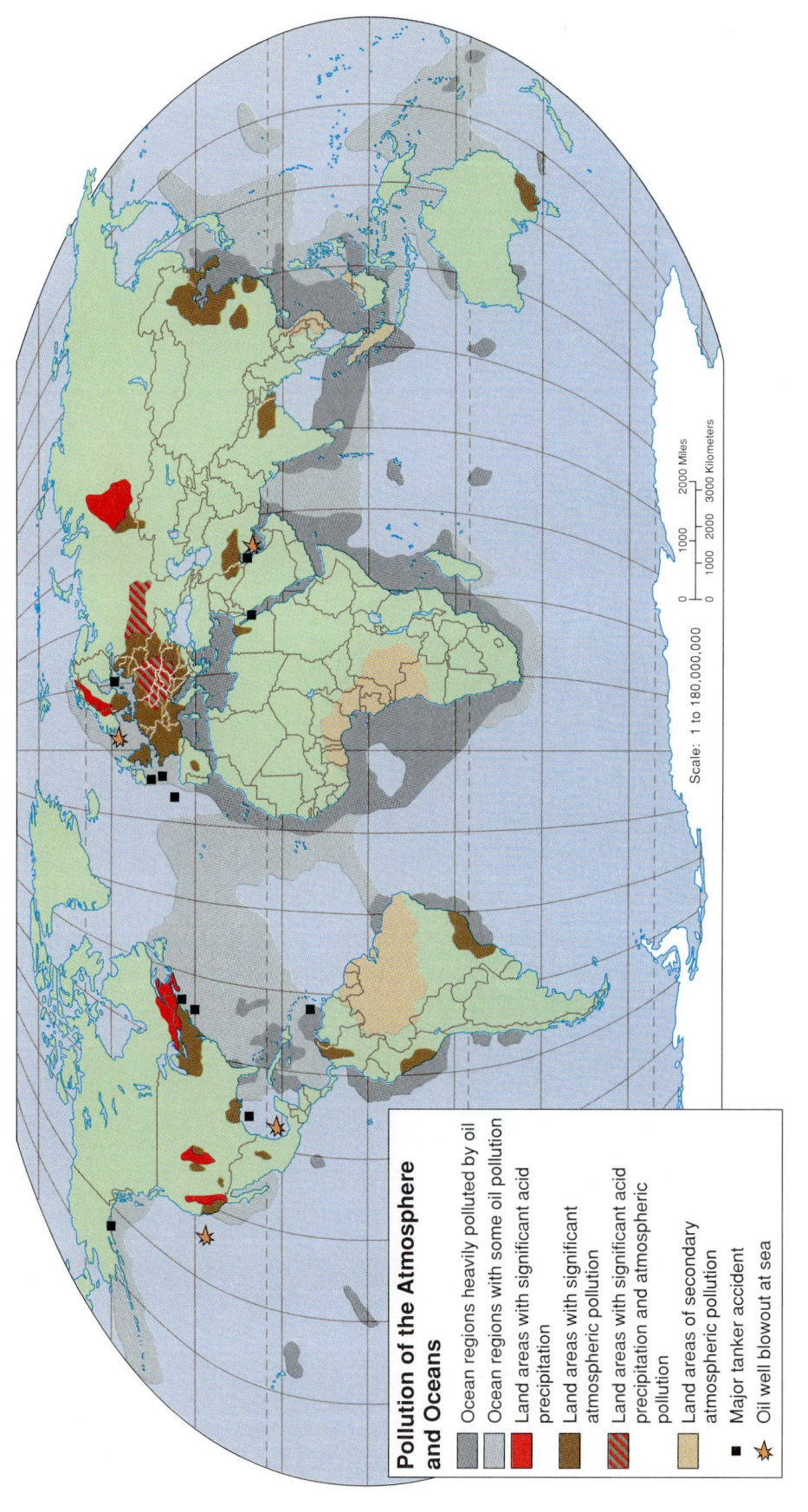

Pollution of the Atmosphere and Oceans

- Ocean regions heavily polluted by oil
- Ocean regions with some oil pollution
- Land areas with significant acid precipitation
- Land areas with significant atmospheric pollution
- Land areas with significant acid precipitation and atmospheric pollution
- Land areas of secondary atmospheric pollution
- ■ Major tanker accident
- ✦ Oil well blowout at sea

Scale: 1 to 180,000,000

The pollution of the world's oceans and atmosphere has long been a matter of concern to environmental scientists. The great circulation systems of ocean and air are the controlling factors of the earth's natural environment, and modifications to those systems have unknown consequences. This map is based on what we can measure: (1) areas of oceans where oil pollution has been proven to have inflicted significant damage to ocean ecosystems and lifeforms (including phytoplankton—the oceans' primary food producers, the equivalent of land vegetation); (2) areas of oceans where unusually high concentrations of hydrocarbons from oil spills may have inflicted some damage on the oceans' biota; (3) land areas where the combination of sulphur and nitrogen oxides with atmospheric water vapor has created acid precipitation at sufficiently high levels to have produced significant damage to terrestrial vegetation systems; (4) land areas where the emissions from industrial, transportation, commercial, residential, and other uses of fossil fuels have produced concentrations of atmospheric pollutants high enough to be damaging to human health; and (5) land areas of secondary air pollution where the primary pollutant is smoke from forest clearance. A glance at the map shows that there are few areas of the world where some form of oceanic or atmospheric pollution is not a part of our environmental system. Scientists are still debating the long-range implications of this pollution, but nearly all agree that the consequences, whatever they may be, will not be good.

– 88 –

Map 45 Per Capita CO_2 Emissions

Per Capita CO_2 Emissions
In metric tons

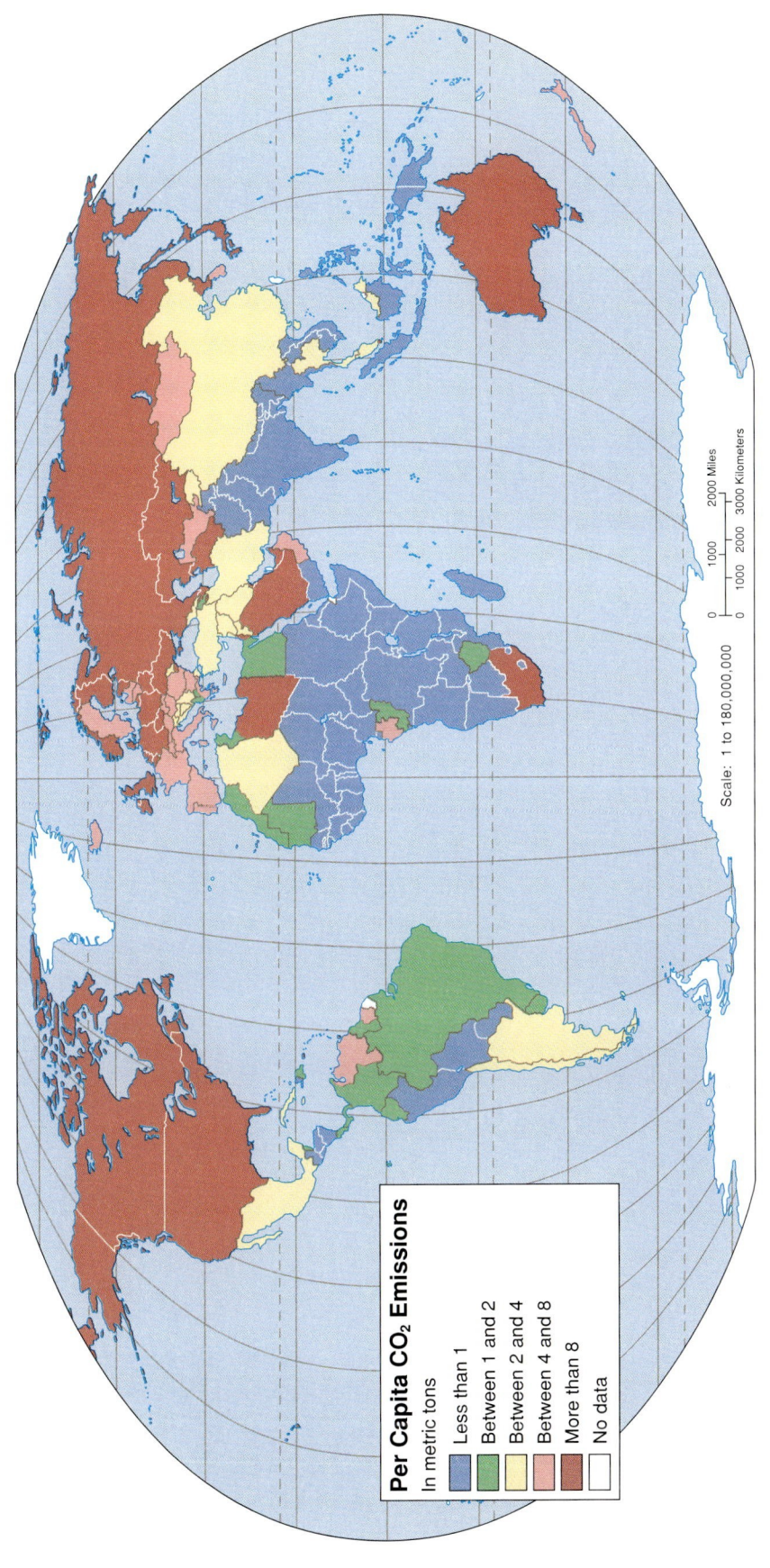

- Less than 1
- Between 1 and 2
- Between 2 and 4
- Between 4 and 8
- More than 8
- No data

Scale: 1 to 180,000,000

Carbon dioxide emissions are a major indicator of economic development, since they are generated largely by the burning of fossil fuels for electrical power generation, for industrial processes, for domestic and commercial heating, and for the internal combustion engines of automobiles, trucks, buses, planes, and trains. Scientists have long known that carbon dioxide in the atmosphere increases the ability of the atmosphere to retain heat, a phenomenon known as the greenhouse effect. While the greenhouse effect is a natural process (and life on earth as we know it would not be possible without it), many scientists are concerned that the increase in carbon dioxide in the atmosphere will augment this process, creating a global warming trend and a potential worldwide change of climate patterns. These climatological changes threaten disaster for many regions and their peoples in both the developed and the less developed areas of the world. You will note from the map that the countries of the Northern Hemisphere generate extremely high levels of carbon dioxide per capita.

Table J
Land Use (in thousands of hectares)

	DOMESTICATED LAND AS A % OF LAND AREA OF COUNTRY	CROPLAND 1991–1993	% CHANGE SINCE 1981–1983[1]	PERMANENT PASTURE 1991–1993	% CHANGE SINCE 1981–1983[1]	FOREST AND WOODLAND 1991–1993	% CHANGE SINCE 1981–1983[1]	OTHER LAND 1991–1993	% CHANGE SINCE 1981–1983[1]
WORLD	38	1,450,834	1.3	3,364,537	3.5	4,168,956	(3.6)	4,114,077	(0.5)
AFRICA	**35**	**187,357**	**5.8**	**853,049**	**1.2**	**760,576**	**(3.1)**	**1,162,630**	**(0.3)**
Algeria	16	7,938	7.0	30,752	(3.2)	3,969	(9.5)	195,516	(0.5)
Angola	26	3,483	2.5	29,000	0.0	51,917	(3.1)	40,270	(3.0)
Benin	21	1,877	4.1	442	0.0	3,407	(12.0)	5,337	(7.3)
Botswana	46	420	5.0	25,600	0.0	26,500	0.0	4,153	0.5
Burkina Faso	35	3,564	23.5	6,000	0.0	13,800	0.0	3,996	17.0
Burundi	89	1,357	3.9	915	0.5	85	0.0	211	26.3
Cameroon	19	7,033	1.2	2,000	0.0	35,900	0.0	1,607	5.2
Central African Republic	8	2,015	2.9	3,000	0.0	46,700	0.0	10,583	0.5
Chad	38	3,239	2.8	45,000	0.0	32,400	0.0	45,281	0.2
Congo	30	170	11.6	10,000	0.0	21,120	(0.9)	2,860	(6.4)
Egypt	8	2,760	11.7	4,934	12.7	31	0.0	91,820	0.9
Equatorial Guinea	12	230	0.0	104	X	1,297	0.1	1,174	0.1
Eritrea	60	1,280	X	1,600	X	2,000	X	5,220	93.5
Ethiopia	53	13,930	0.0	44,825	(1.0)	26,950	(3.4)	24,395	(5.8)
Gabon	20	459	1.5	4,700	0.0	19,900	(0.5)	708	(12.7)
Gambia	27	180	13.0	90	0.0	280	0.0	450	4.6
Ghana	41	4,320	23.4	5,000	0.0	7,943	(8.0)	5,491	2.4
Guinea	25	730	2.4	5,500	0.0	14,480	(3.9)	3,862	(14.6)
Guinea-Bissau	50	340	13.7	1,080	0.0	1,070	0.0	322	12.7
Ivory Coast	53	3,703	16.7	13,000	0.0	7,080	(24.4)	8,017	(21.9)
Kenya	45	4,517	5.5	21,300	0.0	16,800	0.0	14,297	1.7
Lesotho	76	320	10.4	2,000	0.0	80	(4.8)	635	4.1
Liberia	63	374	0.9	5,700	0.0	1,707	(13.2)	1,894	(13.6)
Libya	9	2,167	3.5	13,300	0.8	840	35.1	159,647	0.2
Madagascar	47	3,104	3.3	24,000	0.0	23,200	0.0	7,850	1.3
Malawi	38	1,697	22.1	1,840	0.0	3,700	(1.1)	2,171	12.3
Mali	27	2,270	10.6	30,000	0.0	6,907	(3.9)	82,843	(0.1)
Mauritania	38	207	6.2	39,250	0.0	4,413	(2.1)	58,652	(0.1)
Mauritius	56	106	(0.9)	7	0.0	44	(24.1)	46	(32.6)
Morocco	69	9,781	15.9	20,900	0.0	8,290	6.0	5,659	32.0
Mozambique	60	3,163	2.7	44,000	0.0	14,053	(7.7)	17,192	(6.3)
Namibia	47	662	0.5	38,000	0.0	18,030	(1.8)	25,637	(1.3)
Niger	10	3,605	1.5	8,913	(3.4)	2,500	0.8	111,652	(0.2)
Nigeria	79	32,368	6.1	40,000	0.0	11,400	(20.3)	7,309	(14.1)
Rwanda	66	1,167	8.5	453	(15.2)	550	(4.8)	297	(5.8)
Senegal	28	2,350	0.0	3,100	0.0	10,467	(4.4)	3,336	(14.5)
Sierra Leone	38	540	5.6	2,203	(0.1)	2,043	(2.9)	2,376	(1.4)
Somalia	70	1,032	2.2	43,000	0.0	16,000	6.7	2,702	37.8

(Continued on next page)

	DOMESTICATED LAND AS A % OF LAND AREA OF COUNTRY	CROPLAND 1991–1993	CROPLAND % CHANGE SINCE 1981–1983[1]	PERMANENT PASTURE 1991–1993	PERMANENT PASTURE % CHANGE SINCE 1981–1983[1]	FOREST AND WOODLAND 1991–1993	FOREST AND WOODLAND % CHANGE SINCE 1981–1983[1]	OTHER LAND 1991–1993	OTHER LAND % CHANGE SINCE 1981–1983[1]
South Africa	77	13,177	(0.1)	81,378	(0.0)	8,200	0.0	19,349	(0.1)
Sudan	52	12,950	3.6	110,000	12.2	44,340	(6.1)	70,310	13.6
Swaziland	73	191	35.4	1,070	(6.5)	118	15.7	341	(2.5)
Tanzania	44	3,500	19.2	35,000	0.0	33,500	(14.4)	16,359	(31.0)
Togo	48	2,430	3.0	200	0.0	933	(8.5)	1,876	(0.9)
Tunisia	52	4,897	2.2	3,525	5.1	664	18.8	6,451	5.9
Uganda	43	6,763	13.4	1,800	0.0	5,503	(7.7)	5,898	5.8
Zaire	10	7,893	2.9	15,000	0.0	173,860	(1.7)	29,952	(9.6)
Zambia	47	5,271	2.2	30,000	0.0	28,727	(2.3)	10,341	(5.5)
Zimbabwe	20	2,864	3.4	4,856	0.0	8,800	(7.4)	22,165	(2.7)
NORTH AND MIDDLE AMERICA	**29**	**271,300**	**(0.9)**	**362,033**	**0.1**	**854,897**	**5.7**	**689,945**	**6.3**
Belize	5	57	7.5	48	9.1	2,100	0.0	75	10.7
Canada	8	45,523	(1.3)	27,933	(3.88)	494,000	11.6	354,640	14.0
Costa Rica	56	530	3.5	2,337	9.1	1,570	(5.2)	670	19.1
Cuba	57	3,337	3.8	2,970	11.4	2,403	(9.2)	2,273	8.0
Dominican Republic	30	1,449	1.4	2	0.0	608	(3.7)	2,780	(0.1)
El Salvador	65	731	0.8	610	0.0	104	(18.8)	527	(2.9)
Guatemala	41	1,817	2.3	2,534	91.8	5,271	17.4	1,221	166.5
Haiti	51	908	1.3	495	(2.3)	140	0.0	1,212	(0.0)
Honduras	32	1,904	7.7	1,511	0.7	6,000	0.0	1,774	8.3
Jamaica	44	219	(4.8)	257	0.0	184	(4.5)	423	(4.7)
Mexico	52	24,727	0.2	74,499	0.0	48,700	4.1	42,943	4.6
Nicaragua	57	1,272	1.0	5,483	9.7	3,223	(24.3)	1,896	(28.4)
Panama	29	658	25.0	1,487	10.9	3,260	(18.0)	2,038	(23.8)
Trinidad and Tobago	26	121	3.4	11	0.0	235	3.1	146	7.6
United States	45	187,776	(1.2)	239,172	(0.4)	286,400	(2.0)	243,963	(3.8)
SOUTH AMERICA	**34**	**104,567**	**1.6**	**495,884**	**3.9**	**846,721**	**(4.1)**	**305,753**	**(5.2)**
Argentina	62	27,200	0.0	142,033	(0.7)	50,900	0.0	53,536	(1.8)
Bolivia	27	2,373	11.1	26,517	(1.7)	58,000	0.0	21,549	(1.1)
Brazil	28	50,560	0.3	185,767	6.8	488,833	(4.80)	120,491	(10.3)
Chile	24	4,293	0.3	13,583	3.7	16,500	0.0	40,504	1.2
Colombia	44	5,450	4.2	40,567	4.8	49,533	(5.8)	8,220	(12.1)
Ecuador	18	3,012	20.2	2,091	(0.0)	15,600	0.6	6,981	8.7
Guyana	9	496	0.2	1,230	0.5	16,413	0.3	1,546	3.3
Paraguay	60	2,258	19.3	21,600	30.6	12,983	(32.5)	2,888	(28.6)
Peru	24	3,630	0.7	27,120	0.0	84,800	(0.1)	12,450	(0.2)
Suriname	1	68	24.4	21	6.8	15,000	0.8	511	25.4
Uruguay	85	1,304	(6.8)	13,520	(0.6)	930	0.0	1,727	(10.5)
Venezuela	25	3,912	4.1	17,783	2.8	29,828	(8.2)	36,682	(5.5)
ASIA	**X**	**470,322**	**2.9**	**799,881**	**12.9**	**533,087**	**(1.9)**	**958,375**	**9.1**
Afghanistan	58	8,054	0.0	30,000	0.0	1,900	0.0	25,255	0.0

(Continued on next page)

	DOMESTICATED LAND AS A % OF LAND AREA OF COUNTRY	CROPLAND 1991–1993	CROPLAND % CHANGE SINCE 1981–1983[1]	PERMANENT PASTURE 1991–1993	PERMANENT PASTURE % CHANGE SINCE 1981–1983[1]	FOREST AND WOODLAND 1991–1993	FOREST AND WOODLAND % CHANGE SINCE 1981–1983[1]	OTHER LAND 1991–1993	OTHER LAND % CHANGE SINCE 1981–1983[1]
Armenia	X	X	X	X	X	X	X	2,840	0.0
Azerbaijan	49	2,000	2.0	2,233	(4.1)	960	(12.5)	3,417	(5.7)
Bangladesh	79	9,703	6.1	600	0.0	1,896	(12.2)	818	36.3
Bhutan	9	133	6.7	272	2.5	3,100	20.4	1,194	45.3
Cambodia	25	2,367	47.9	1,967	239.1	11,667	(11.3)	1,652	40.0
China	X	95,975	(3.3)	X	X	X	X	836,666	(0.4)
Georgia	43	993	0.3	2,033	(9.8)	2,717	(6.1)	1,227	(32.2)
India	61	169,547	0.6	11,533	(4.1)	68,330	1.4	47,909	3.0
Indonesia	24	30,993	19.2	11,776	(1.3)	111,258	(3.7)	27,130	2.1
Iran	38	18,057	22.1	44,000	0.0	11,400	0.0	90,143	3.6
Iraq	22	5,450	0.1	4,000	0.0	192	0.0	34,095	0.0
Israel	28	435	4.8	145	12.7	124	12.7	1,358	3.7
Japan	14	4,511	(6.6)	656	9.8	25,187	(0.0)	7,299	(3.7)
Jordan	13	404	18	791	0.1	70	6.6	7,628	0.9
Kazakhstan	83	35,328	(1.6)	186,452	(1.0)	9,600	(6.5)	35,600	(8.6)
Korea, North	17	2,010	4.5	50	0.0	2	0.0	9,979	0.9
Korea, South	22	2,072	(4.9)	91	43.9	7,370	0.0	340	(23.1)
Kuwait	8	5	150.0	137	2.0	6,464	(1.4)	(4,824)	1.8
Kyrgyzstan	54	1,387	(3.9)	8,943	(1.7)	2	0.0	8,798	(2.4)
Laos	7	807	11.6	800	0.0	703	(11.3)	20,770	(0.0)
Lebanon	31	306	2.7	10	0.0	2,819	2.3	(2112)	(3.4)
Malaysia	15	4,880	0.4	27	0.0	20,347	(1.8)	7,601	(4.6)
Mongolia	81	1,399	11.1	124,800	1.1	13,750	(9.4)	16,701	0.6
Myanmar	16	10,061	(0.1)	359	(0.7)	32,397	0.9	22,938	1.2
Nepal	32	2,354	1.5	2,000	4.2	5,750	4.4	3,576	9.9
Oman	5	62	49.6	1,000	0.0	X	X	20,184	0.1
Pakistan	34	22,890	12.4	5,000	0.0	3,470	14.5	45,728	6.5
Philippines	35	9,177	3.8	1,277	17.1	13,600	13.4	5,764	37.0
Saudi Arabia	58	3,719	75.7	120,000	41.2	1,800	36.7	89,450	41.5
Singapore	X	1	(84.2)	X	X	3	0.0	57	(9.4)
Sri Lanka	36	1,903	2.1	439	0.1	2,126	21.4	1,995	20.7
Syria	75	5,770	0.8	8,018	(4.0)	679	37.9	3,911	(2.5)
Tajikistan	31	836	(7.2)	3,507	1.1	535	7.8	9,391	0.1
Thailand	42	20,775	9.4	797	17.2	13,557	(13.8)	15,960	(1.7)
Turkey	52	27,583	0.6	12,378	22.6	20,199	0.0	16,803	14.6
Turkmenistan	77	1,462	(39.1)	36,274	0.8	4,000	(10.4)	7,074	(16.0)
United Arab Emirates	3	39	36.0	200	0.0	3	0.0	8,118	0.1
Uzbekistan	63	4,728	4.6	22,183	(5.9)	1,323	(42.5)	14,306	(15.1)
Vietnam	22	6,607	0.4	328	7.9	9,639	(8.8)	15,975	(5.5)
Yemen	33	1,481	1.1	16,065	0.0	2,000	(33.3)	33,251	(3.0)
EUROPE	**X**	**136,757**	**(2.7)**	**80,794**	**5.6**	**158,219**	**1.4**	**808,204**	**0.2**
Albania	41	702	(0.8)	424	4.7	1,050	2.3	564	6.6
Austria	42	1,509	(5.6)	1,978	(2.2)	3,228	(1.0)	1,557	(10.8)
Belarus	45	6,255	(1.7)	3,128	(6.2)	7,009	(5.2)	4,369	(16.0)

(Continued on next page)

	DOMESTICATED LAND AS A % OF LAND AREA OF COUNTRY	CROPLAND 1991–1993	CROPLAND % CHANGE SINCE 1981–1983[1]	PERMANENT PASTURE 1991–1993	PERMANENT PASTURE % CHANGE SINCE 1981–1983[1]	FOREST AND WOODLAND 1991–1993	FOREST AND WOODLAND % CHANGE SINCE 1981–1983[1]	OTHER LAND 1991–1993	OTHER LAND % CHANGE SINCE 1981–1983[1]
Belgium	52	1,007	30.6	691	(9.4)	700	0.4	885	(18.9)
Bosnia-Herzegovina	38	1,007	X	1,067	X	2,100	X	927	450.4
Bulgaria	55	4,267	2.8	1,878	(7.4)	3,875	0.5	1,035	(1.5)
Croatia	43	1,415	(13.2)	1,247	(21.2)	2,081	2.3	849	(59.5)
Czech Republic	54	3,293	X	873	X	2,629	X	933	728.3
Denmark	65	2,549	(3.3)	206	(14.9)	445	(9.7)	1,043	(16.3)
Estonia	34	1,146	15.2	312	(6.8)	1,986	19.2	783	57.2
Finland	9	2,560	2.1	116	(23.6)	23,198	(0.5)	4,587	(2.3)
France	55	19,297	1.6	11,023	(12.8)	14,884	2.0	9,806	(10.4)
Germany	50	12,009	(3.7)	5,274	(11.0)	10,700	4.0	6,943	(10.1)
Greece	68	3,506	(11.2)	5,253	(0.0)	2,620	0.0	1,510	(29.4)
Hungary	66	5,077	(4.2)	1,165	(9.2)	1,726	6.1	1,266	(19.2)
Iceland	23	6	(20.8)	2,274	0.0	120	0.0	7,625	(0.0)
Ireland	81	926	(12.2)	4,691	0.5	320	(0.8)	951	(11.1)
Italy	55	11,915	(3.7)	4,284	(15.7)	6,769	6.2	6,438	(13.2)
Latvia	41	1,715	(1.3)	822	9.0	12,507	(5.6)	(8,839)	7.8
Lithuania	77	3,043	(3.6)	459	(15.1)	1,980	1.0	(930)	18.9
Macedonia	51	663	X	636	X	1,002	X	242	952.3
Moldova	79	2,202	(0.4)	378	8.0	421	74.3	296	67.5
Netherlands	59	922	12.0	1,065	(10.1)	343	15.9	1,061	2.4
Norway	3	888	5.8	120	18.4	8,330	0.0	21,344	0.3
Poland	61	14,694	(0.9)	4,043	(0.6)	8,779	0.9	2,926	(2.7)
Portugal	44	3,169	0.7	839	0.1	3,293	12.3	1,894	20.2
Romania	64	9,974	(5.3)	4,820	8.7	6,681	1.8	1,559	(3.5)
Russia	12	133,141	(1.8)	77,985	(6.8)	778,512	1.9	709,941	0.9
Serbia-Montenegro (Yugoslavia)	X	7,730	(1.4)	6,352	(0.6)	9,120	(1.4)	(23,202)	1.2
Slovakia	51	1,623	X	825	X	1,990	X	370	1,198.3
Slovenia	43	302	X	559	X	1,018	X	133	1,412.8
Spain	60	19,897	(2.9)	10,281	(2.7)	15,970	2.4	3,796	(13.6)
Sweden	8	2,779	(5.8)	573	(18.9)	28,000	0.1	9,809	(2.9)
Switzerland	40	449	9.1	1,279	(20.5)	1,185	12.7	1,043	(15.3)
Ukraine	72	34,458	(2.9)	7,471	6.6	10,278	36.4	5,728	37.9
United Kingdom	71	6,442	(7.7)	11,112	(1.6)	2,424	12.0	4,182	(11.0)
OCEANIA	**57**	**51,619**	**4.3**	**430,738**	**(3.7)**	**199,962**	**24.4**	**163,030**	**15.2**
Australia	60	46,579	3.8	416,567	(3.7)	145,000	36.8	156,298	15.9
Fiji	24	257	42.8	174	34.1	1,185	0.0	211	57.6
New Zealand	65	3,831	9.4	13,577	(5.3)	7,377	3.6	2,014	(8.3)
Papua New Guinea	1	412	11.1	81	(17.3)	42,000	0.0	2,793	0.9
Solomon Islands	3	57	8.2	39	0.0	2,450	(3.5)	253	(33.9)

[1] Negative numbers are shown in parentheses.

Sources: Food and Agriculture Organization of the United Nations; United Nations Population Division; *World Resources 1996–97*.

Table K
Human-Induced Soil Destruction, 1945–1990

	TOTAL DEGRADED AREA[1] (millions of hectares)	DEGRADED AREA AS % OF VEGETATED LAND[2]	EROSION[3] (millions of hectares) WATER	WIND	DEGRADATION[4] (millions of hectares) CHEMICAL	PHYSICAL
WORLD	1,964.4	17	1,093.7	548.3	239.1	83.3
Light Degradation	749.0	6	343.2	268.6	93.0	44.2
Moderate Degradation	910.5	8	526.7	253.6	103.3	26.8
Strong Degradation	295.7	3	217.2	24.3	41.9	12.3
Extreme Degradation	9.3	0	6.6	1.9	0.8	0.0
AFRICA	494.2	22	227.4	186.5	61.5	18.7
Light Degradation	173.6	8	57.5	88.3	26.0	1.8
Moderate Degradation	191.8	9	67.4	89.3	27.0	8.1
Strong Degradation	123.6	6	98.3	7.9	8.6	8.8
Extreme Degradation	5.2	0	4.2	1.0	0.0	0.0
NORTH AND MIDDLE AMERICA	158.1	8	106.1	39.2	7.0	5.9
Light Degradation	18.9	1	14.5	2.6	0.5	1.3
Moderate Degradation	112.5	5	68.2	34.9	5.7	3.8
Strong Degradation	26.7	1	23.4	1.7	0.8	0.8
Extreme Degradation	0.0	0	0.0	0.0	0.0	0.0
SOUTH AMERICA	243.4	14	123.2	41.9	70.3	7.9
Light Degradation	104.8	6	45.9	25.8	26.3	6.8
Moderate Degradation	113.5	7	65.1	16.1	31.4	0.8
Strong Degradation	25.0	1	12.1	0.0	12.6	0.3
Extreme Degradation	0.0	0	0.0	0.0	0.0	0.0
ASIA[5]	748.0	20	440.6	222.2	73.2	12.1
Light Degradation	294.5	8	124.5	132.4	31.8	5.7
Moderate Degradation	344.3	9	241.7	75.1	21.5	6.0
Strong Degradation	107.7	3	73.4	14.5	19.5	0.4
Extreme Degradation	0.5	0	0.0	0.2	0.4	0.0
EUROPE[5]	218.9	23	114.5	42.2	25.8	36.4
Light Degradation	60.6	6	21.4	3.2	8.1	27.9
Moderate Degradation	144.4	15	81.0	38.2	17.1	8.1
Strong Degradation	10.7	1	9.8	0.0	0.6	0.4
Extreme Degradation	3.1	0	2.4	0.7	0.0	0.0
OCEANIA	102.9	13	82.8	16.4	1.3	2.3
Light Degradation	96.6	12	79.4	16.3	0.2	0.7
Moderate Degradation	3.9	0	3.2	0.0	0.7	0.0
Strong Degradation	1.9	0	0.2	0.1	0.0	1.6
Extreme Degradation	0.4	0	0.0	0.0	0.4	0.0

[1]Degraded soils are those in which the current and/or future capacity to produce agricultural output has been lowered.

[2]Vegetated land includes all natural vegetated areas and all agricultural land.

[3]Erosion is topsoil loss, the removal of soil by wind action, surface wash, or sheet erosion.

[4]Degradation includes both chemical and physical soil deterioration. Chemical degradation includes nutrient loss from overcultivation, salinization from irrigation, pollution from agricultural chemicals and other sources, and acidification from overfertilization or acid precipitation. Physical degradation includes compaction from livestock or heavy machinery, waterlogging through overirrigation or drainage disturbances, and subsidence by drainage disturbance or oxidation.

[5]The values for Europe include Belarus, Moldava, Ukraine, Armenia, Azerbaijan, Georgia, and Russia west of the Ural Mountains. The values for Asia include Russia east of the Urals, along with Kazakhstan, Uzbekistan, Tadjikistan, Turkmenistan, and Kyrgyzstan.

Sources: United Nations Environmental Program; International Soil Reference and Information Center; *World Resources 1996–97.*

Table L
Water Resources, 1995

	ANNUAL RENEWABLE WATER RESOURCES[1]		ANNUAL WITHDRAWALS[2]		
	TOTAL (KM^3)	PER CAPITA (M^3)	TOTAL (KM^3)	% OF RESOURCES	PER CAPITA (M^3)
WORLD	41,022.0	7,176	3,240.00	8	645
AFRICA	3,996.0	5,488	145.14	4	199
Algeria	14.8	528	4.5	30	180
Angola	184.0	16,618	0.48	0	57
Benin	25.8	4,770	0.15	1	28
Botswana	14.7	9,886	0.11	1	83
Burkina Faso	28.0	2,713	0.38	1	40
Burundi	3.6	563	0.10	3	20
Cameroon	268.0	20,252	0.40	0	38
Central African Republic	141.0	42,534	0.07	0	26
Chad	43.0	6,760	0.18	0	34
Congo	832.0	321,236	0.04	0	20
Egypt	58.1	923	56.40	97	956
Equatorial Guinea	30.0	75,000	0.01	0	15
Eritrea	8.8	2,492	X	X	X
Ethiopia	110.0	1,998	2.21	2	51
Gabon	164.0	124,242	0.06	0	57
Gambia	8.0	7,156	0.02	0	30
Ghana	53.2	3,048	0.30	1	35
Guinea	22.6	33,731	0.74	0	140
Guinea-Bissau	27.0	25,163	0.02	0	17
Ivory Coast	77.7	5,451	0.71	1	66
Kenya	30.2	1,069	2.05	7	87
Lesotho	5.2	2,551	0.05	1	30
Liberia	232.0	76,341	0.13	0	56
Libya	0.6	111	4.60	767	880
Madagascar	337.0	22,827	16.30	5	1,584
Malawi	18.7	1,678	0.94	5	86
Mali	67.0	6,207	1.36	2	162
Mauritania	11.4	5,013	1.63	14	923
Mauritius	2.2	1,979	0.36	16	410
Morocco	30.0	1,110	10.85	36	427
Mozambique	208.0	12,997	0.61	0	41
Namibia	45.5	29,545	0.25	1	180
Niger	32.5	3,552	0.50	2	69
Nigeria	280.0	2,506	3.63	1	41
Rwanda	6.3	792	0.77	12	102
Senegal	39.4	4,740	1.36	3	202
Sierra Leone	160.0	35,485	0.37	0	99
Somalia	13.5	1,459	0.81	6	98
South Africa	50.0	1,206	13.31	27	359
Sudan	154.0	5,481	17.80	12	633
Swaziland	4.5	5,275	0.66	15	1,171
Tanzania	89.0	2,998	1.16	1	40
Togo	12.0	2,900	0.09	1	28
Tunisia	3.9	443	3.08	78	381
Uganda	66.0	3,009	0.20	0	20
Zaire	1,019.0	23,211	0.36	0	10
Zambia	116.0	12,267	1.71	1	186
Zimbabwe	20.0	1,776	1.22	6	136
NORTH AND MIDDLE AMERICA	6,443.7	15,359	608.44	9	1,451
Belize	16.0	74,419	0.02	0	109
Canada	1,901.0	98,462	45.10	2	1,602
Costa Rica	95.0	27,745	1.35	1	780
Cuba	34.5	3,125	8.10	23	870
Dominican Republic	20.0	2,557	2.97	15	446
El Salvador	19.0	3,285	1.00	5	245
Guatemala	116.0	10,922	0.73	1	139

(Continued on next page)

	ANNUAL RENEWABLE WATER RESOURCES[1]		ANNUAL WITHDRAWALS[2]		
	TOTAL (KM^3)	PER CAPITA (M^3)	TOTAL (KM^3)	% OF RESOURCES	PER CAPITA (M^3)
Haiti	11.0	1,532	0.04	0	7
Honduras	63.4	11,216	1.52	2	294
Jamaica	8.3	3,392	0.32	4	159
Mexico	357.4	3,815	77.62	22	899
Nicaragua	175.0	39,477	0.89	1	367
Panama	144.0	54,732	1.30	1	754
Trinidad and Tobago	5.1	3,905	0.15	3	148
United States	2,478.0	9,413	467.34	19	1,870
SOUTH AMERICA	**9,526.00**	**29,788**	**106.21**	**1**	**332**
Argentina	994.0	28,739	27.60	4	1,043
Bolivia	300.0	40,464	1.24	0	201
Brazil	6,950.0	42,957	36.47	1	246
Chile	468.0	32,814	16.80	4	1,626
Colombia	1,070.0	30,483	5.34	0	174
Ecuador	314.0	27,400	5.56	2	581
Guyana	241.0	288,623	1.46	1	1,812
Paraguay	314.0	63,306	0.43	0	109
Peru	40.0	1,682	6.10	15	300
Suriname	200.0	472,813	0.46	0	1,189
Uruguay	124.0	38,920	0.65	1	241
Venezuela	1,317.0	60,291	4.10	0	382
ASIA	**13,206.7**	**3,819**	**1,633.85**	**12**	**542**
Afghanistan	50.0	2,482	26.11	52	1,830
Armenia	13.3	3,687	3.80	46	1,145
Azerbaijan	33.0	4,364	15.80	56	2,248
Bangladesh	2,357.0	19,571	22.50	1	220
Bhutan	95.0	57,998	0.02	0	14
Cambodia	498.1	48,590	0.52	0	64
China	2,800.0	2,292	460.00	16	461
Georgia	65.2	11,942	4.00	7	741
India	2,085.0	2,228	380.00	18	612
Indonesia	2,530.0	12,804	16.59	1	96
Iran	117.5	1,746	45.40	39	1,362
Iraq	109.2	5,340	42.80	43	4,575
Israel	2.2	382	1.85	86	408
Japan	547.0	4,373	90.80	17	735
Jordan	1.7	314	0.45	32	173
Kazakhstan	169.4	9,900	37.90	30	2,294
Korea, North	67.0	2,801	14.16	21	687
Korea, South	66.1	1,469	27.60	42	632
Kuwait	0.2	103	27.60	X	525
Kyrgyzstan	61.7	13,003	11.70	24	2,729
Laos	270.0	55,305	0.99	0	259
Lebanon	5.6	1,854	0.75	16	271
Malaysia	456.0	22,642	9.42	2	768
Mongolia	24.6	10,207	0.44	2	273
Myanmar	1,082.0	23,255	3.96	0	101
Nepal	170.0	7,756	2.68	2	150
Oman	1.9	892	0.48	24	564
Pakistan	468.0	3,331	153.40	33	2,053
Philippines	323.0	4,770	29.50	9	686
Saudi Arabia	4.6	254	3.60	164	497
Singapore	0.6	211	0.19	32	84
Sri Lanka	43.2	2,354	6.30	15	503
Syria	53.7	3,662	3.34	9	435
Tajikistan	101.3	16,604	12.80	13	2,455
Thailand	179.0	3,045	31.90	18	602
Turkey	193.10	3,117	33.50	17	585
Turkmenistan	72.0	17,573	22.80	33	6,390
United Arab Emirates	2.0	1,047	0.90	299	884
Uzbekistan	129.6	5,674	82.20	76	4,121

(Continued on next page)

	ANNUAL RENEWABLE WATER RESOURCES[1]		ANNUAL WITHDRAWALS[2]		
	TOTAL (KM^3)	PER CAPITA (M^3)	TOTAL (KM^3)	% OF RESOURCES	PER CAPITA (M^3)
Vietnam	376.0	5,044	28.90	8	414
Yemen	5.2	359	3.40	136	335
EUROPE	**6,234.5**	**8,576**	**455.29**	**7**	**626**
Albania	21.3	6,190	0.20	1	94
Austria	90.3	11,333	2.36	3	304
Belgium	73.8	7,277	3.00	5	295
Bosnia-Herzegovina	X	X	X	X	X
Bulgaria	12.5	1,236	9.03	72	917
Croatia	61.4	13,660	X	0	X
Czech Republic	58.2	5,653	2.74	5	266
Denmark	13.0	2,509	1.20	9	233
Estonia	17.6	11,490	3.30	21	2,097
Finland	113.0	22,126	2.20	2	440
France	198.0	3,145	37.73	19	665
Germany	171.0	2,096	46.27	27	579
Greece	58.7	5,612	5.04	9	523
Hungary	120.0	11,864	6.81	6	661
Iceland	168.0	624,535	0.16	0	636
Ireland	50.0	14,073	0.79	2	233
Italy	167.0	2,920	56.20	34	986
Latvia	34.0	13,297	0.70	2	262
Lithuania	24.2	6,541	4.40	19	1,190
Macedonia	X	X	X	X	X
Moldova	13.7	3,093	3.70	29	853
Netherlands	90.0	5,805	7.81	9	518
Norway	392.0	90,385	2.03	1	488
Poland	56.2	1,464	12.28	22	321
Portugal	69.6	7,085	7.29	10	739
Romania	208.0	9,109	26.00	13	1,134
Russia	4,498.0	30,599	117.00	3	790
Serbia-Montenegro (Yugoslavia)	X	X	X	X	X
Slovakia	30.8	5,753	1.78	6	337
Slovenia	X	X	X	X	X
Spain	111.3	2,809	30.75	28	781
Sweden	180.0	20,501	2.93	2	341
Switzerland	50.0	6,943	1.19	2	173
Ukraine	231.0	4,496	34.70	40	673
United Kingdom	71.0	1,219	11.79	17	205
OCEANIA	**1,614.3**	**56,543**	**16.73**	**1**	**586**
Australia	343.0	18,693	14.60	4	933
Fiji	28.6	36,416	0.03	0	42
New Zealand	327.0	91,469	2.00	1	589
Papua New Guinea	801.0	186,192	0.10	0	28
Solomon Islands	44.7	118,254	0.00	0	0

[1] Annual internal renewable water resources refer to the average annual flow of rivers plus annual groundwater recharge from precipitation.

[2] Annual withdrawals refer to the total use of water, not including evaporation losses from storage reservoirs and soil and transpiration loss from vegetation. Water withdrawals also include water from desalinization plants in countries where that use is a significant part of all water withdrawals.

Sources: Bureau of Geological Research and Mining, National Geological Survey, France; Institute of Geography, National Academy of Sciences, Russia; Eurostat; International Desalinization Association; United Nations Population Bureau; *World Almanac and Book of Facts 1996; World Development Report 1995; World Resources 1996–97.*

Table M
Greenhouse Gas Emissions, 1992[1] (in thousands of metric tons)

	INDUSTRIAL CO$_2$[2]	LAND USE CO$_2$[3]	METHANE[4]	PER CAPITA CO$_2$
WORLD	22,339,408	4,100,000	270,00	4.10
AFRICA	715,773	730,000	21,000	X
Algeria	79,712	6,900	1,700	3.00
Angola	4,525	16,000	340	0.44
Benin	612	3,200	65	0.11
Botswana	2,173	3,200	110	1.65
Burkina Faso	557	3,400	260	0.07
Burundi	191	130	33	0.04
Cameroon	2,231	28,000	260	0.18
Central African Republic	216	23,000	110	0.07
Chad	253	7,100	240	0.04
Congo	3,972	14,000	31	1.69
Egypt	83,997	X	1,000	1.54
Equatorial Guinea	117	3,000	2	0.33
Eritrea	X	X	68	X
Ethiopia	2,906	8,000	1,200	0.04
Gabon	5,569	51,000	250	4.51
Gambia	198	130	20	0.22
Ghana	3,781	18,000	130	0.22
Guinea	1,026	10,000	520	0.18
Guinea-Bissau	209	1,800	59	0.22
Ivory Coast	6,309	15,000	110	0.48
Kenya	5,342	1,400	540	0.22
Lesotho	X	X	44	X
Liberia	278	9,600	31	0.11
Libya	39,520	76	480	8.10
Madagascar	945	21,000	860	0.07
Malawi	652	11,000	72	0.07
Mali	443	8,400	350	0.04
Mauritania	2,869	1	140	1.36
Mauritius	1,356	9	5	1.25
Morocco	27,344	4,700	360	1.03
Mozambique	997	15,000	98	0.07
Namibia	X	1,800	96	X
Niger	1,085	X	160	0.15
Nigeria	96,513	24,000	4,500	0.84
Rwanda	451	170	34	0.07
Senegal	2,810	4,700	190	0.37
Sierra Leone	432	1,800	150	0.11
Somalia	15	430	540	0.00
South Africa	290,291	14,000	2,400	7.29
Sudan	3,462	38,000	1,100	0.15
Swaziland	267	370	25	0.33
Tanzania	2,103	22,000	760	0.07
Togo	733	2,100	45	0.18
Tunisia	13,560	640	130	1.61
Uganda	953	5,000	250	0.04
Zaire	4,181	280,000	380	0.11
Zambia	2,481	34,000	150	0.29
Zimbabwe	18,675	5,300	230	1.76
NORTH AND MIDDLE AMERICA	5,715,466	190,000	35,000	X
Belize	264	980	4	1.32
Canada	409,862	X	3,500	14.99
Costa Rica	3,807	14,000	130	1.21
Cuba	28,623	3,200	370	2.64
Dominican Republic	10,248	4,800	210	1.36
El Salvador	3,550	520	88	0.66
Guatemala	5,657	21,000	160	0.59
Haiti	784	470	88	0.11
Honduras	3,059	19,000	130	0.55
Jamaica	8,042	7,200	31	3.26
Mexico	332,852	63,000	3,100	3.77

(Continued on next page)

	INDUSTRIAL CO$_2$[2]	LAND USE CO$_2$[3]	METHANE[4]	PER CAPITA CO$_2$
Nicaragua	2,495	33,000	140	0.82
Panama	4,228	21,000	100	1.69
Trinidad and Tobango	20,643	1,200	300	16.30
United States	4,881,349	X	27,000	19.13
SOUTH AMERICA	**605,029**	**1,800,000**	**140,000**	**X**
Argentina	117,003	85,000	3,700	3.52
Bolivia	6,632	140,000	540	0.88
Brazil	217,074	1,100,000	9,900	1.39
Chile	34,738	33,000	380	2.56
Colombia	61,493	110,000	2,100	1.83
Ecuador	18,888	72,000	490	1.72
Guyana	835	6,900	31	1.03
Paraguay	2,620	35,000	440	0.59
Peru	22,277	96,000	520	0.99
Suriname	2,008	5,100	42	4.58
Uruguay	5,038	1,300	690	1.61
Venezuela	116,424	170,000	2,000	5.75
ASIA	**7,118,317**	**1,300,000**	**140,000**	**X**
Afghanistan	1,392	1,100	210	0.07
Armenia	4,199	X	42	1.21
Azerbaijan	63,878	X	450	8.76
Bangladesh	17,217	7,700	3,900	0.15
Bhutan	132	4,500	36	0.07
Cambodia	476	35,000	140	0.04
China	2,667,982	150,000	47,000	2.27
Georgia	13,839	X	130	2.53
India	769,440	65,000	33,000	0.88
Indonesia	184,585	410,000	10,000	0.95
Iran	235,478	10,000	3,300	3.81
Iraq	64,527	24	540	3.33
Israel	41,605	X	73	8.10
Japan	1,093,470	X	3,900	8.79
Jordan	11,311	97	50	2.64
Kazakhstan	297,982	X	2,500	17.48
Korea, North	252,750	700	1,600	11.21
Korea, South	289,833	1,500	1,400	6.56
Kuwait	15,971	X	170	8.10
Kyrgyzstan	X	X	190	X
Laos	271	X	270	0.07
Lebanon	11,051	91	29	3.88
Malaysia	70,492	210,000	960	3.74
Mongolia	9,281	480	300	4.03
Myanmar	4,386	130,000	2,300	0.11
Nepal	1,297	9,000	610	0.07
Oman	10,036	X	150	6.12
Pakistan	71,902	14,000	3,300	0.59
Phillippines	49,698	110,000	1,900	0.77
Saudi Arabia	220,620	60	2,600	13.85
Singapore	49,790	X	65	17.99
Sri Lanka	4,972	4,300	610	0.29
Syria	42,407	790	620	3.19
Tajikistan	3,972	X	150	0.70
Thailand	112,477	92,000	5,500	2.02
Turkey	145,490	X	1,100	2.49
Turkmenistan	42,257	X	1,000	10.96
United Arab Emirates	70,616	X	520	42.28
Uzbekistan	123,253	X	1,300	5.75
Vietnam	21,522	40,000	4,400	0.29
Yemen	10,083	X	90	0.81
EUROPE	**6,866,494**	**11,000**	**53,000**	**X**
Albania	3,968	X	92	1.21
Austria	56,572	X	260	7.29
Belarus	102,028	X	510	9.89
Belgium	101,768	X	190	10.19
Bosnia-Herzegovina	15,055	X	38	3.37

(Continued on next page)

	INDUSTRIAL CO$_2$[2]	LAND USE CO$_2$[3]	METHANE[4]	PER CAPITA CO$_2$
Bulgaria	54,359	X	8,800	6.08
Croatia	16,210	X	73	3.33
Czech Republic	135,608	X	380	13.04
Denmark	53,897	X	260	10.44
Estonia	20,885	X	59	13.19
Finland	41,176	X	150	8.21
France	362,076	X	1,800	6.34
Germany	878,136	X	3,400	10.96
Greece	73,859	X	330	7.25
Hungary	59,910	X	270	5.72
Iceland	1,777	X	15	6.85
Ireland	30,851	X	520	8.87
Italy	407,701	X	1,500	7.03
Latvia	14,781	X	100	5.53
Lithuania	22,006	X	150	5.86
Macedonia	4,100	X	40	1.98
Moldova	14,209	X	110	3.26
Netherlands	139,027	X	1,400	9.16
Norway	60,247	X	2,400	14.03
Poland	341,892	X	1,800	8.90
Portugal	47,181	X	170	4.80
Romania	122,103	X	840	5.24
Russia	2,103,132	X	17,000	14.11
Serbia-Montenegro (Yugoslavia)	38,197	X	180	3.63
Slovakia	36,999	X	320	7
Slovenia	5,503	X	35	2.75
Spain	223,196	X	1,400	5.72
Sweden	56,796	X	200	6.56
Switzerland	43,701	X	140	6.38
Ukraine	611,342	X	3,600	11.72
United Kingdom	566,246	X	3,800	9.78
OCEANIA	**297,246**	**38,000**	**5,800**	**X**
Australia	267,937	X	4,800	15.24
Fiji	711	1,400	25	0.95
New Zealand	26,179	X	1,000	7.58
Papua New Guinea	2,257	35,000	13	0.55
Solomon Islands	161	1,800	1	0.48

[1]Greenhouse gases are those gases, occurring either naturally or through human activities, that enhance the ability of the earth's atmosphere to trap and retain heat energy. Heat energy is solar energy that has been absorbed by the earth's land and water surfaces, converted from light energy to long-wave or heat energy, and radiated back to warm the atmosphere. Many atmospheric scientists believe that an increase in the atmospheric content of greenhouse gases through fossil-fuel burning, forest clearance, and other anthropogenic processes may cause global warming.

[2]Industrial carbon dioxide results from the combustion of solid, liquid, and gas fuels, gas flaring during petroleum extraction, and cement manufacture.

[3]Land use carbon dioxide generation is produced by land use changes that create higher than normal emissions; chief among these is forest clearance by burning, but the category would also include wetlands restoration, irrigation agriculture, and livestock feeding.

[4]Methane (CH$_4$) is produced chiefly from oil and natural gas extraction and distribution, coal mining, wetland rice agriculture, municipal solid waste decomposition, and livestock.

Source: World Resources 1996–97.

Sources

Crabb, C. (1993, January). Soiling the planet. *Discover, 14* (1), 74-75. [For information regarding Map 42 on page 86 of this book]

DeBlij, H. J., & Muller, P. (1994). *Geography: Realms, regions and concepts* (7th ed., revised). New York: John Wiley & Sons.

Domke, K. (1988). *War and the changing global system.* New Haven, CT: Yale University Press.

Information please almanac 1996. (1995). Boston & New York: Houghton Mifflin.

Johnson, D. (1977). *Population, society, and desertification.* New York: United Nations Conference on Desertification, United Nations Environment Programme.

Köppen, W., & Geiger, R. (1954). *Klima der erde* [Climate of the earth]. Darmstadt, Germany: Justus Perthes.

Lindeman, M. (1990). *The United States and the Soviet Union: Choices for the 21st century.* Guilford, CT: Dushkin Publishing Group.

Murphy, R. E. (1968). Landforms of the world [Map supplement No. 9]. *Annals of the Association of American Geographers, 58* (1), 198-200.

National Oceanic and Atmospheric Administration. (1990-1992). Unpublished data. Washington, DC: NOAA.

Population Reference Bureau. (1994). *1994 world population data sheet.* New York: Population Reference Bureau.

Rourke, J. T. (1995). *International politics on the world stage* (5th ed.). Guilford, CT: Dushkin Publishing Group/Brown & Benchmark Publishers.

Spector, L. S., & Smith, J. R. (1990). *Nuclear ambitions: The spread of nuclear weapons.* Boulder, CO: Westview Press.

United Nations Population Fund. (1992). *The state of the world's population.* New York: United Nations Population Fund.

U.S. Arms Control and Disarmament Agency. (1993). *World military expenditures and arms transfers.* Washington, DC: U.S. Government Printing Office.

U.S. Census Bureau. (1994). *World population profile.* Washington, DC: U.S. Government Printing Office.

U.S. Forest Service. (1989). *Ecoregions of the continents.* Washington, DC: U.S. Government Printing Office.

The world almanac and book of facts 1996. (1995). Mahwah, NJ: World Almanac Books.

World Bank. (1995). *World development report 1995.* Geneva: World Bank.

World Health Organization. (1994). *World health statistics annual.* Geneva: World Health Organization.

World Resources Institute. (1996). *World resources 1996-97.* New York: Oxford University Press.